Soviet and Russian Lunar Exploration

Brian Harvey

Soviet and Russian Lunar Exploration

 Springer

Published in association with
Praxis Publishing
Chichester, UK

Brian Harvey
2 Rathdown Crescent
Terenure
Dublin 6W
Ireland

SPRINGER–PRAXIS BOOKS IN SPACE EXPLORATION
SUBJECT *ADVISORY EDITOR*: John Mason, M.Sc., B.Sc., Ph.D.

ISBN 10: 0-387-21896-3 Springer Berlin Heidelberg New York
ISBN 13: 978-0-387-21896-0 Springer Berlin Heidelberg New York

Springer is part of Springer-Science + Business Media (springer.com)

Library of Congress Control Number: 2006935327

Cover design: Jim Wilkie
Project management: Originator Publishing Services, Gt Yarmouth, Norfolk, UK

Printed on acid-free paper

Contents

Acknowledgements

The author would like to thank and acknowledge all those who assisted with this book. In particular, he would like to thank: Dave Shayler, whose creative ideas helped to shape this book; Rex Hall, for his comments and advice on the Soviet cosmonaut squad and making available his collection for study; Phil Clark of Molniya Space Consultancy, for his technical advice; Paolo Ulivi, Bart Hendrickx and Don P. Mitchell who provided valuable information; Andy Salmon, for access to his collection; Suszann Parry, for making available information sources and photographs in the British Interplanetary Society; Prof. Evert Meurs, director and Carol Woods, librarian of Dunsink Observatory; and of course Clive Horwood for his support for this project.

Many of the photographs published here come from the author's collection. I would like to thank the many people who generously provided or gave permission for the use of photographs, especially the following:

- Piet Smolders, for permission to use his painting of the first Russian on the moon.
- Andy Salmon, for his series of images of the LK and Luna 10.
- Rex Hall, for his photographs of Luna 10, 13, 16 and Lunokhod 3.
- NASA, for its collection on Soviet space science. Other open American sources were used, such as the declassified CIA collection.

Brian Harvey
Dublin, Ireland, 2007

Prologue

Siberia, summer of 1976. Near the lowland town of Surgut on the River Ob in western Siberia, Russia's space recovery forces had gathered to await the return of Russia's latest moon probe. Already, the short, warm and fly-ridden Siberian summer was passing. Although it was only the 21st August, the birch trees were already turning colour and there was a cool breeze in the evening air. Gathered on the ground were amphibious army vehicles, designed to carry troops across marshy or rough terrain. In the air were half a dozen Mil helicopters, ready to spot a parachute opening in the sky. Getting to the moon probe quickly was important. They had missed Luna 20 four years earlier: it had come down, unseen, on an island in the middle of a snow-covered river, but thankfully they found it before the battery of its beeping beacon had given out. The diesel engines of the army ground crews were already running. The army crews stood around, waiting, waiting.

Bang! There was the sharp echo of a small sonic boom as the spherical spacecraft came through the sound barrier 20 km high. By this stage, it had barrelled through the high atmosphere at a speed of 7 km/sec, hitting the spot on the tiny 10-km by 20-km entry corridor necessary to ensure a safe return to Earth. The heatshield glowed red, then orange, then white hot as the cabin shed speed for heat. On board, in a sealed container, were precious rock and soil granules drilled up from the distant Sea of Crises on the moon's northeastern face. The probe had left the moon three days earlier. Now, through the most perilous phase of the return, the cabin dropped, unaided, through the ever-denser layers of Earth's atmosphere.

Fifteen kilometres high above the marshes, a meter sensed the growing density of Earth's air. The lid of the cabin was explosively blown off. A small drogue parachute fluttered out. At 11 km, it had pulled out a much larger red-and-white canopy, ballooning out above the still-steaming sphere. Two beacons popped out of the top of the cabin. Abruptly halted in its downward spiral, the cabin twisted and was now caught in the wind and began to drift sideways and downward. The helicopter crews spotted the cabin in the air and picked up the beacon on their radios.

Over their radiophones they called up the amphibians who headed straight in the direction of the returning spacecraft. The helicopters saw the cabin reach the ground. The small parachute at once emptied and deflated to lie alongside. In minutes the amphibians had drawn up alongside. The army crews cut the parachute free. Gingerly – it was still warm from the hot fires of reentry – they lifted the blackened cabin into the back of their vehicle, driving back into Surgut. Within hours, it was on its way by air to the Moscow Vernadsky Institute. This was the third set of samples the Soviet Union had brought back from the moon. The first had come from the Sea of Fertility in 1970, with Luna 16. Two years later, Luna 20 had brought back a small sample from the Apollonius Highland. Luna 24 had gone a stage further and drilled deep into the lunar surface and this cabin had the deepest, biggest sample of moon soil of them all.

Nobody realized at the time that this was the last lunar mission of the Soviet Union. Fifty years later, lunar exploration is remembered for who won, the United States and who lost, the Soviet Union. In the popular mind, the view is that the Russians just did not have the technological capacity to send people to the moon. In reality, political rather than technical reasons prevented the Soviet Union from landing cosmonauts on the moon. It is often forgotten that the story of Soviet lunar exploration is, although it had its fair share of disappointments, also one of achievement. The Soviet Union:

- Sent the first spacecraft past the moon (the First Cosmic Ship).
- Launched the first spacecraft to impact on the lunar surface (the Second Cosmic Ship).
- Sent the first spacecraft around the farside of the moon to take photographs (the Automatic Interplanetary Station).
- Made the first soft-landing on the moon (Luna 9).
- Put the first orbiter into lunar orbit (Luna 10).
- Pioneered sophisticated, precise high-speed reentries into the Earth's atmosphere from the moon, becoming the first country to send a spaceship around the moon and recover it on Earth (Zond 5).
- Landed advanced roving laboratories that explored the moon for months on end (the Lunokhods).
- Retrieved two sets of rock samples from the surface of the moon by automatic spacecraft (Luna 16, 20) and drilled into the surface for a core sample (Luna 24).
- Returned a substantial volume of science from its lunar exploration programme.

Not only that, but the Soviet Union:

- Came close to sending a cosmonaut around the moon first.
- Built and successfully tested, in orbit, a lunar lander, the LK.
- Built a manned lunar orbiter, the LOK.
- Assembled and trained a team of cosmonauts to explore the moon's surface, even selecting sites where they would land.
- Came close to perfecting a giant moon rocket, the N-1.
- Designed long-term lunar bases.

Although the United States Apollo programme is one of the great stories of human-kind, the story of Soviet and Russian lunar exploration is one worth telling too. First designs for lunar exploration date to the dark, final days of Stalin. The Soviet Union mapped out a plan for a lunar landing and, in pursuit of this, achieved most of the key 'firsts' of lunar exploration. Even when the manned programme faltered, a credible programme of unmanned lunar exploration was carried out, one which Luna 24 brought to an end. The story of Soviet lunar exploration is one of triumph and heartbreak, scientific achievement, engineering creativity, treachery and intrigue. Now, new lunar nations like China and India are following in the paths mapped out in the Soviet Union 60 years ago. And Russia itself is preparing to return to the moon, with the new Luna Glob mission in planning.

Figures

Tables and maps

TABLES

MAPS

Abbreviations and acronyms

AIS	Automatic Interplanetary Station
AKA	Also Known As
BOZ	*Blok Obespecheyna Zapushka* (Ignition Insurance System)
B, V, G, D	Letters in Russian alphabet
ΔV	Velocity change
EOR	Earth Orbit Rendezvous
EVA	Extra Vehicular Activity
GDL	Gas Dynamics Laboratory
GIRD	Group for the study of jet propulsion
GNP	Gross National Product
GSLV	Geo Stationary Launch Vehicle
HF and LF	High Frequency and Low Frequency
HTP	High Test Peroxide
ICBM	InterContinental Ballistic Missile
IKI	Institute for Space Research
IZMIRAN	Institute of Terrestrial Magnetism
KBOM	*Konstruktorskoye Buro Obshchevo Mashinostroeniya* (General Engineering Design Bureau)
KL-1E	'E' for Experimental
KORD	*KOntrol Roboti Dvigvateli* (Engine Control System)
KTDU-5	*Korrektiruiushaya Tormoznaya Dvigatelnaya Ustanovka* (Engine Correction System)
KVD	*Kislorodno Vodorodni Dvigatel* (oxygen hydrogen engine)
L	*Luna, Luniy* (moon)
L-1P	'P' for Preliminary
L-1S	'S' for Simplified
L-3M	'M' for Modified
LEK	Lunar expeditionary craft; Lunar Exploration Council

LK	*Luna Korabl*; *Luniy Korabl* (lunar spacecraft)
LM	Lunar Module (Apollo)
LOI	Lunar Orbit Insertion
LOK	*Luniy Orbitalny Korabl* (moon orbital craft)
LOR	Lunar Orbit Rendezvous
LOX	Liquid OXygen
LZhM	Lunar habitation module
LZM	Laboratory production module
N-1	*Nositel* (carrier) 1
NK	Nikolai Kuznetsov, engine
NKVD	Soviet internal security police
NPO	*Nauchno Proizvodstvennoe Obedinenie* (Research Production Association)
NSSDC	National Space Science Center
OB	Cocooned habitation block
OK	*Orbitalny Korabl* (orbital craft)
OKB-1	*Opytno Konstruktorskoye Buro* (Experimental Design Bureau)
os	Old Style, calendar used before the October Revolution, running 12 days behind the rest of Europe
PDI	Powered Descent Initiation
PrOP	*Pribori Ochenki Prokhodimosti* (Terrain Evaluation Instrument), penetrometer
RIFMA	Roentgen Isotopic Fluorescent Method of Analysis
RLA	Rocket Launch Apparatus
SKB 2	*Spetsealnoye Konstruktorskoye Buro 2* (Special Design Bureau 2)
SLV	Satellite Launch Vehicle
TEI	Trans-Earth Injection
TMK	Heavy interplanetary ship
TsDUC	Centre for Long Range Space Communications
TsKBEM	Central Design Bureau of Experimental Machine Building
TsKBM	Central Design Bureau of Machine Building
TsPK	Centre for Cosmonaut Training
UDMH	Unsymmetrical dimethyl methyl hydrazine

1

Origins of the Soviet lunar programme

The Soviet moon programme began in an unlikely place – in a children's magazine, on 2nd October 1951. Mikhail Tikhonravov was a veteran rocket engineer from the 1920s and was now convinced that a flight to the moon might soon become a practical possibility. In the paranoia of Stalin's Russia, talking about unapproved projects like moon flights was a potentially dangerous enterprise, so he chose a relatively safe outlet, one unlikely to raise the blood pressure of the censors: the pages of *Pionerskaya Pravda*, the newspaper devoted to communist youth. There, on 2nd October 1951, he outlined how two men could fly out to the moon and back in a 1,000 tonne rocketship. The article concluded:

We do not have long to wait. We can assume that the bold dream of Tsiolkovsky will be realized within the next 10 to 15 years. All of you will become witness to this and some of you may even be participants in unprecedented journeys.

His article was noticed immediately by Western intelligence, which apparently scanned children's magazines as well the main national political press. In what may have been the first occasion that Soviet space plans were noticed in the West, the *New York Times* noted 'Dr Tikhonravov's article', commenting that Soviet advances in rockets were developing rapidly and might equal, if not exceed, Western achievements. Indeed, at official level within the Soviet Union, his article was noticed too, for when the next edition of the *Great Soviet Encyclopaedia* came to be written, Mikhail Tikhonravov was invited to write a section called *Interplanetary communications* (1954) [1].

The next step took place in April 1954, a year after the death of Stalin. Although there was no direct connexion between scientific research institute NII-4 (NII stands for Scientific Research Institute, or in Russian *Nauchno Issledovatelsky Institut*), where Mikhail Tikhonravov was posted and the OKB-1 experimental design bureau (in Russian, *Opytno Konstrucktorskoye Buro*), where the chief designer of spaceflight

Mikhail Tikhonravov

Sergei Korolev worked, there was clearly a degree of informal collaboration between them. In 1946, Stalin had appointed a council of spaceflight designers and it was headed by a 'chief designer' (in Russian *Glavnykonstruktor*). The chief designer was Sergei Korolev, the legend who led the Soviet space programme from its inception. The chief designer was not just a crucial engineering post, but the political leader of the space programme, making it the most coveted position in the industry. His support was now critical.

May 1954 was the deadline for proposals for projects for countries interested in participating in the forthcoming International Geophysical Year. Encouraged, indeed prompted by Sergei Korolev, the Russian proposal was written by Mikhail Tikhonravov, in consultation with leading Soviet mathematician Mstislav Keldysh and Russia's top rocket engine designer, Valentin Glushko. Called *Report on an artificial satellite of the Earth*, it was, according to historian Siddiqi, one of the great researchers of the period, a *tour de force* of foresight for the 1950s and remarkable even in the present day [2]. Even though the Soviet Union had yet to commit itself to a small Earth satellite, the writers tried to engage their country in a project for manned spaceflight from the very start. The third section of the report dealt with the problems of reaching the moon and outlined how the rocket that they were then building could send a probe to the moon and bring it back to Earth through means of atmospheric braking. *Report on an artificial satellite of the Earth* did not emerge from the archives until the 1990s, but it was the first mention, in an official document of plans for a Soviet flight to the moon. Although the report appeared at first sight to sink in a sea of red bureaucratic ink, in fact it became the basis of the Soviet space programme. Siddiqi says that the combination of Korolev's managerial genius and Tikhonravov's technical acumen became the basis of humankind's departure from the Earth.

Sergei Korolev, Mstislav Keldysh

With the Soviet Union at last thawing out from the time of terror, it was now possible to discuss lunar missions more openly. The 25th September 1955 marked the 125th anniversary of the NE Baumann Moscow Higher Technical School. Here, chief designer Sergei Korolev gave a lengthy paper called *On the question of the application of rockets for research into the upper layers of the atmosphere*. Here, he outlined the possibility of landing robotic probes on the surface of the moon. As the chief designer, Korolev had developed a series of rockets, derived from the German V-2, firing some with animals into the upper atmosphere. Now under Soviet Premier Nikita Khrushchev he was tasked with developing the Soviet Union's first intercontinental ballistic missile (ICBM), capable of hitting the United States. The postwar Soviet rocket effort was driven by two complementary imperatives. The political leadership wanted missiles, while the engineers wanted rockets to explore space. Engineers had to justify their rocket building in terms of their military capability and potential. Only later did the political leadership appreciate that missiles designed for military purposes could also be powerful servants of non-military political objectives. While the intercontinental ballistic missile would indeed, Korolev knew, meet Khrushchev's military needs, Korolev always designed the rocket with a second purpose in mind: to open the door to space travel.

THE 1956 LENINGRAD CONFERENCE

The following year, the State University of Leningrad convened a conference of physicists to examine the nature of the moon and the planets. It was held in Leningrad in February 1956. Most of those attending were scientists, astronomers and what would now be called planetologists. Also there was Mikhail Tikhonravov, not representing *Pionerskaya Pravda*, but this time the Artillery Institute, where NII-4 was located. The conference in Leningrad State University, which reviewed the state of knowledge of our moon at the time, was well publicized and news of its deliberations were again picked up in the West [3].

Following the deliberations in Leningrad State University, Korolev paid a visit to Tikhonravov's Artillery Institute. There, he asked its designers, engineers and experts to explain their work to him. As was his wont, Korolev said little, preferring to listen and taking a particular interest in their work on trajectories. Being a man more of action than of words, the institute soon found out that it had made its mark. Wielding his authority as chief designer, Korolev transferred the institute to his own, the first experimental design bureau, OKB-1. There, the NII-4 personnel could be under his direct control and enlisted fully in his cause. They now became department #9 of OKB-1, founded 8th March 1957 [4]. We do not know what Mikhail Tikhonravov thought of this. He was a quiet man who preferred to work in the background and who rarely sought the limelight. His unassuming nature concealed great imagination, a steely sense of purpose and, as the situation in the early 1950s required some considerable courage.

This was typical of Korolev. Long before his intercontinental ballistic missile had flown, some time before the first Sputnik had even been approved, he was already thinking ahead to a flight to the moon. Working on several projects at once daunted many lesser men, but it was his *forte*. Korolev's drive, imagination, timing and ability to knock heads (and institutes) together do much to explain the early successes of the Soviet space programme [5]. The relationship between Tikhonravov and Korolev has attracted little attention, but it was a key element in the early Soviet lunar programme. One person who has commented is Sergei Khrushchev, son of the Soviet leader Nikita Khrushchev. Sergei Khrushchev says that Korolev was not an originator of technical ideas, but someone able to gather the best engineers and technicians around him. He was able, though, to spot talent, to organize, to manage, to drive ideas and concepts through the political system. Although many of the ideas of his design bureau were attributed externally to him, he made sure that, within the bureau, individual designers were recognized, promoted, praised and rewarded. Khrushchev: 'Mikhail Tikhonravov was a man of brilliant intellect and imaginative scope [but] totally devoid of organizational talents' [6]. The combination of Korolev the organizer and Tikhonravov the designer worked well and between them they built the moon programme.

INTRODUCING THE FATHER OF THE SOVIET MOON PROGRAMME: MIKHAIL TIKHONRAVOV

Tikhonravov's background in the space programme went as far back as Korolev's, even though he was much less publicly prominent. But what do we know about Mikhail Tikhonravov? Mikhail Tikhonravov was the architect of the Soviet moon

GIRD-09

programme. He was born 16th July 1900 (os)[1] and began his early aeronautical career by studying the flight characteristics of birds and insects. In 1922, his study called *Some statistical and aerodynamical data on birds* was published in *Aircraft* magazine. He graduated from the Zhukovsky air force academy in 1925 and worked in aviation. In 1932 he joined Korolev's group of amateur rocketeers, the GIRD (Group for the study of jet propulsion), moving in and out of rocketry and jet propulsion in the 1930s and 1940s. He wrote *Density of air and its change with altitude* for a military magazine in 1924. Seven more articles on aeronautics appeared by 1939. In the course of his work he met the ageing theoretician Konstantin Tsiolkovsky and joined the Moscow GIRD. He was closely involved with Korolev in the construction of amateur rockets

[1] os is Old Style, the calendar in use before the Bolshevik revolution, which ran twelve days behind the rest of Europe. New style dates are given for those born after the revolution.

launched over 1933–5. The Moscow group had fired the first liquid-fuel Russian rocket from a forest near Moscow. The rocket was called the GIRD-09, a needle-like contraption just able to fly higher than the tall trees. Launching on 17th August 1933, it reached the mighty height of 400 m in its 18 sec mission. The GIRD rocket was designed by Mikhail Tikhonravov. The work of these young rocketeers and theore-ticians was later to become extremely significant for the later moon missions. GIRD was supervised by a technical council with four teams, led respectively by Friedrich Tsander, Sergei Korolev, Yuri Pobedonostsev and Mikhail Tikhonravov, with Tikhonravov having responsibility for liquid propellants [7]. The group was really driven by Sergei Korolev (born 30th December 1906 (os)), a graduate of Moscow Higher Technical School who designed, built and flew his own gliders and for which he developed rockets as a means to get them airborne.

Tikhonravov wrote a book on space travel in 1935 and then disappears from the records until the end period of the war. He was one of the few to escape the purges. Tikhonravov was a talented man who painted oils in his spare time and studied insects and beetles. Tikhonravov re-emerged in 1944 designing high-altitude rockets for the Lebedev Institute of the Academy of Sciences and two years later was transferred to Scientific Research Institute NII-4, staffed mainly by artillery officers, to design and build missiles. In the later 1940s, his name reappears on an edited book on the writings of Konstantin Tsiolkovsky and Friedrich Tsander. Tikhonravov designed the first plans for sending humans into space – the VR-190 suborbital rocket, able to send two stratonauts on an up-and-down mission 200 km high, a flight eventually emulated by Alan Shepard and Virgil Grissom in 1961. From 1948 onward, Tikhonravov worked for the Artillery Academy of Sciences and put forward the idea of grouping rockets together in a cluster of packets to achieve new velocities and lifting power. It was at such a presentation attended by Korolev in 1948 that the two men resumed their collaboration that had been broken by the purges [8]. On 15th March 1950, Tikhon-ravov put forward one of the formative papers of the Soviet space programme, with a convoluted but self-explanatory title: *On the possibility of achieving first cosmic veloc-ity and creating an Earth satellite with the aid of a multi-stage missile using the current level of technology.*

This paper caused a stir and indeed led to Tikhonravov's banishment. In the final, paranoid days of Stalin, he fell under suspicion for giving unwarranted attention to non-military affairs and for not concentrating sufficiently on the defence of the motherland. He was demoted, rather than imprisoned or worse, but ironically this gave him all the more time to consider long-term objectives. During this period of reflection, the article for *Pionerskaya Pravda* was conceived. Following the death of Stalin, he was restored to his old work in the Directorate of the Deputy Commander of Artillery. There, he organized the 'satellite team' that paved the way for the Soviet Union to launch the first Sputnik. His memorandum *A report on an artificial satellite of the Earth* (25th May 1954) included a final section called *Problems of reaching the moon* which outlined a 1,500 kg spacecraft to land on the moon and return using atmospheric braking. His ideas had now moved from a children's newspaper to an official Soviet document in the period of three years.

April 1956 saw the Soviet Academy of Sciences organize the all-Union conference

On rocket research into the upper layers of the atmosphere. Here, Sergei Korolev made a lengthy presentation. He told the conference:

It is also a real task to prepare the flight of a rocket to the moon and back to the Earth. The simplest way to solve this problem is to launch a probe from an Earth satellite orbit. At the same time, it is possible to perform such a flight directly from the Earth. These are prospects of the not too distant future.

Department #9 was later reorganized and subtitled the 'Planning department for the development of space apparatus'. In April 1957, the planning department produced a detailed technical document, *A project research plan for the creation of piloted satellites and automatic spacecraft for lunar exploration.* The key question, iterated by Tikhonravov, was the need to construct an upper stage for the planned intercontinental ballistic missile. Meantime, the Academy of Sciences appointed the Commission on Interplanetary Communications to oversee the planning or 'the conquest of cosmic space': vice-chairman was Mikhail Tikhonravov.

There the matter rested for the moment, as OKB-1 focused on the great challenge of launching an artificial Earth satellite that autumn.

Chronology of the idea of a Soviet moon rocket
1951   *Flight to the moon* by Mikhail Tikhonravov in *Pionerskaya Pravda.*
1954   *Report on an artificial satellite of the Earth* by Tikhonravov, Glushko and Keldysh.
1955   *On the question of the application of rockets for research into the upper layers of the atmosphere* by Sergei Korolev.
1956   Conference on moon in Leningrad State University (February).
        Korolev formally announces goal of moon mission (April) at conference *On rocket research into the upper layers of the atmosphere.*
        Artillery institute's research institute NII-4 transferred to OKB-1 as Department #9 under Tikhonravov.
1957   Department #9's *Project research plan for the creation of piloted satellites and automatic spacecraft for lunar exploration* (April).
        Academy of Sciences establishes the Commission on Interplanetary Communications, led by Tikhonravov.

SOVIET SPACE PROGRAMME BEFORE SPUTNIK

The Soviet space programme before Sputnik was the coming together of a number of diverse bodies, people, institutes and traditions. Going to the moon, Earth's nearest celestial neighbour, had always been a part of this idea.

The Soviet space programme actually stretched back into Tsarist times. Its chief visionary was a deaf schoolteacher, Konstantin Tsiolkovsky (1857–1935). He was a remarkable man who carried out space experiments in his home, drew designs for interstellar spacecraft, calculated rocket trajectories (Tsiolkovsky's formula is still taught in mathematics) and wrote science fiction about the exploration of the solar system. Rocketry was little encouraged under the tsars – indeed, another early designer, Nikolai Kibalchich, was executed in 1881 for turning his knowledge of explosives to use in an assassination plot.

The 1920s became the golden age of theoretical Soviet cosmonautics. Popular societies blossomed, exhibitions were held, science fiction was published, an encyclopaedia of space travel issued. It was rich in theoretical, practical and popular work. Friedrich Tsander and Alexander Shargei (AKA Yuri Kondratyuk) outlined how spacecraft could fly to the moon and Mars. Popular societies were set up to popularize space travel and exhibitions were held. In St Petersburg, the Gas Dynamics Laboratory (GDL) was set up in the old St Peter and Paul Fortress. It attracted the brightest Russian chemical engineer of the century, Valentin Glushko and here the first static Russian rocket engines were developed. Glushko, born 20th August 1908 (os), was a precocious young engineer who had built a toy rocket at age 13, corresponded with Tsiolkovsky in 1923 and wrote his own first contributions on spaceflight in 1924. He joined the original rocket engine design bureau in Russia, the Gas Dynamics Laboratory, in 1925 and was given his own subdivision in 1929, when he was just over 20 years old. The following year, Glushko began his first experiments with nitric acid fuels and developed new ways of insulating rocket engines through exotic materials like zirconium. 1931 found him working on self-igniting fuels, swivelling (gimballing) engines and high-speed turbine pumps.

Alexander Shargei addressed some of the key questions of lunar missions in *The conquest of interplanetary space* (1929). He put forward the notion that, in landing on the moon or planets, the landing stage should be left behind and used as a launching

Valentin Glushko, chief designer

pad for the returning spacecraft. He suggested that it would be more economical to land on a moon or planet from an orbit, rather than by a direct descent. He outlined how explorers from the moon and planets could return by using the Earth's atmosphere to break their speed through reentry. In 1930, the elderly Konstantin Tsiolkovsky was advisor to a film called *Kosmicheskoye putechestviye* (Space journey), a Mosfilm spectacular in which spacesuited Soviet cosmonauts travelled weightless to the moon (the actors were suspended on wires to simulate zero gravity) and then walked its surface.

This flourishing of theory, practice and literature came to an abrupt and grotesque end in 1936 with the start of the great purges. The head of the army's rocket programme, Marshal Tukhachevsky, was seized, charged with treason and shot, all within a matter of hours. Sergei Korolev was sent off to the gulag and Glushko was put under arrest for six years. The leaders of GDL, Langemaak and Kleimenov, were shot. Most other engineers were put under house arrest and very few escaped the wrath of Stalin in some shape or form (lucky Tikhonravov was one of them). The amateur societies were closed down. Fortunate was Tsiolkovsky not to see all this, for he died in old age in 1935.

The survivors of the Gulags were let out – or kept under a relaxed form of arrest – to contribute to the war effort. Rocketeers now put their talents to work in aircraft design to win the war against Germany. Their real shock came in 1944 when they learned of the progress made by Germany in rocket design. Mikhail Tikhonravov was one of a team of Russian scientists to visit Poland in August 1944 behind then rapidly retreating German lines. They went there on foot of intelligence reports sent to Britain which indicated that Germany was developing a rocket weapon. Following the RAF attack on the main German launch site at Peenemünde, Germany had moved testing to an experimental station in Debica, Poland, near the city of Krakow. Polish agents had found the launch and impact sites there and had managed to salvage the remains of the rocket, including, crucially, the engine. British prime minister Winston Churchill asked Stalin to facilitate access by British experts to the site, though this meant of course that Stalin's experts would benefit equally from what they found. They found that Germany had stolen a march on them all and under the guidance of their chief designer, Wernher von Braun, had launched the world's first real ballistic rocket, the A-4, on 3rd October 1942. A month after Tikhonravov's visit to Poland, the first A-4s were fired as a military weapon. Over 1944–5, the A-4, renamed the V-2, was used to bombard London and Antwerp. The Germans had also moved ahead with sophisticated guided missiles (like the *Schmetterling*) and anti-aircraft missiles (like the *Wasserfall*) and were far advanced in a range of related technologies. In early 1945, the Red Army swept into the development centre of the A-4, the Baltic launch site of Peenemünde.

THE POSTWAR MOBILIZATION

Neither the Russians, Americans, British nor French were under any misapprehensions about the achievement of von Braun and his colleagues. Each side dispatched its top rocket experts to Germany to pick over the remains of the A-4. For one brief moment in time, all the world's great rocket designers were within a few kilometres of each another. Von Braun was there, though busily trying to exfiltrate himself to America. For the Soviet Union, Valentin Glushko, Sergei Korolev, Vasili Mishin, Georgi Tyulin and Boris Chertok. For the United States, Theodor von Karman, William Pickering and Tsien Hsue Shen (who eventually became the founder of the Chinese space programme). Later in 1945, Britain was to fire three V-2s over the North Sea. Britain's wartime allies were invited to watch. The British admitted one 'Colonel Glushko' but they refused admittance to another 'Captain Korolev' because his paperwork was not in order and he had to watch the launching from the perimeter fence. The British were not fooled by these civilians in military uniforms, for they could give remarkably little account of their frontline experience (or wounds) in the course of four years' warfare.

Korolev and Glushko returned to Russia where Stalin put them quickly to work to build up a Soviet rocket programme. The primary aim was to develop missiles and if the engineers entertained ambitions for using them for space travel, they may not have kept Stalin so fully informed. The rocket effort was reorganized, a series of design bureaus being created from then onwards, the lead one being Korolev's own, OKB-1. Glushko was, naturally, put in charge of engines (OKB-456). In 1946, the Council of Designers was created, Korolev as chief designer. This was a significant development, for it included all the key specialisms necessary for the later lunar programme: engines (Valentin Glushko), radio systems (Mikhail Ryazansky), guidance (Nikolai Pilyugin), construction (Vladimir Barmin) and gyros (Viktor Kuznetsov). In 1947, the Russians managed to fire the first of a number of German A-4s from a missile base, Kapustin Yar, near Stalingrad on the River Volga. The Russian reverse-engineered version was called the R-1 (R for rocket, *Raket* in Russian) and its successors became the basis for the postwar Soviet missile forces. Animals were later launched on up-and-down missions on later derivatives, like the R-5.

The significant breakthrough that made possible the development of space travel was an intercontinental ballistic missile (ICBM). In the early 1950s, as the Cold War intensified, the rival countries attempted to develop the means of delivering a nuclear payload across the world. The ICBM was significant for space travel because the lifting power, thrust and performance required of an ICBM was similar to that required for getting a satellite into orbit. In essence, if you could launch an ICBM, you could launch a satellite. And if you could launch a satellite, you could later send a small payload to the moon.

Approval for a Soviet ICBM was given in 1953. An ICBM in the 1950s was a step beyond the A-4, as much as the A-4 of the 1940s was a step beyond the tiny amateur rockets of the 1930s. Korolev was the mastermind of what became known as the R-7 rocket. It was larger than any rocket built before. It used a fuel mixture of liquid oxygen and kerosene, a significant improvement on the alcohol used on the German A-4. Its powerful engines were designed and built by Valentin Glushko, whose own design bureau, OKB-456, was now fully operational. The real breakthrough for the R-7 was that in addition to the core stage with four engines (block A), four stages of similar dimensions were grouped around its side (blocks B, V, G and D). This was

The R-7

called a 'packet' design – an idea of Mikhail Tikhonravov dating to 1947 when he worked for NII-4. No fewer than 20 engines fired at liftoff. Work began on this project over 1950–3.

The new rocket required a new cosmodrome. Kapustin Yar was too close to American listening bases in Turkey. A new site was selected at Tyuratam, north of a bend in the Syr Darya river, deep in Kazakhstan. The launch site was called Baiko-nour, but this was a deliberate deception. Baikonour was actually a railhead 280 km to the north, but the Russians figured that if they called it Baikonour and if nuclear war broke out, the Americans would mistakenly target their warheads on the small, undefended unfortunate railway station to the north rather than the real rocket base. Construction of the new cosmodrome started in 1955, the labourers living and work-ing in primitive and hostile conditions. Their first task was to construct, out of an old quarry, a launch pad and flame pit. The first pad was built to take the new ICBM, the R-7.

Scientific direction for the space programme was provided by the Academy of Sciences. The Academy dated back to the time of Peter the Great. Following the European tradition, he established a centre of learning for Russia's academic com-munity in St Petersburg. This had survived the revolution, though now it was renamed the Soviet Academy of Sciences. For the political leadership's point of view, the Academy provided a visible and acceptable international face for a space programme that had its roots in military imperatives. The chief expert on the space programme within the Academy of Sciences was Mstislav Keldysh, a quiet, graying, mathematical academician. Mstislav Keldysh was son of Vsevolod M. Keldysh (1878–1965), one of the great engineers of the early Soviet state, the designer of the Moscow Canal, the Moscow Metro and the Dniepr Aluminium Plant. Young Mstislav was professor of aerohydrodynamics in Moscow University, an academician in 1943 at the tender age of 32 and from 1953 director of the Institute of Applied Mathematics. Following Stalin's death, he had introduced computers into Soviet industry. He was on the praesidium of the Academy from 1953, won the Lenin Prize in 1957 and later, from 1961 to 1975, was academy president. He was the most prestigious scientist in the Soviet Union, though he made little of the hundreds of awards with which he was showered in his lifetime. His support and that of the academy for Korolev and Tikhonravov was to become critical.

In the 1950s, the idea of a Russian space programme enjoyed discussion in the popular Soviet media. The golden age of the 1920s had come to an abrupt end in 1936 and talking about space travel remained dangerous as long as Stalin ruled the Kremlin. When the political environment thawed out, ideas around space travel once again flourished in the Soviet media – newspapers, magazines and film. Soviet astronomers resumed studies that had been interrupted by the war. A department of astrobotany was founded by the Kazakh Academy of Sciences and its director, Gavril Tikhov, publicized the possibililities of life on Mars and Venus. His books were wildly popular and he toured the country giving lectures.

By 1957, the key elements of the Russian space programme were in place:

- A strong theoretical base.
- Practical experience of building engines from the 1920s and small rockets from the 1930s.
- A council of designers, led by a chief designer.
- A lead design bureau, OKB-1, with a specialized department, #9.
- Specialized design bureaux for all critical support areas, such as engines.

- An academy of sciences, to provide scientific direction.
- Launch sites in Kapustin Yar and Baikonour.
- Popular and political support.
- A large rocket, completing design.

REFERENCES

[1] Gorin, Peter A: Rising from the cradle – Soviet public perceptions of space flight before Sputnik. From: Roger Launius, John Logsdon and Robert Smith: *Reconsidering Sputnik – 40 years since the Soviet satellite*. Harwood Academic, Amsterdam, 2000.

[2] Siddiqi, Asif: Early satellite studies in the Soviet Union, 1947–57. Part 2. *Spaceflight*, vol. 39, #11, November 1997.

[3] Siddiqi, Asif: The decision to go to the moon. *Spaceflight*,
   – vol. 40, #5, May 1998 (part 1);
   – vol. 40, #6, June 1998 (part 2).

[4] Varfolomeyev, Timothy: Soviet rocketry that conquered space. *Spaceflight*, in 13 parts:
   1 Vol. 37, #8, August 1995;
   2 Vol. 38, #2, February 1996;
   3 Vol. 38, #6, June 1996;
   4 Vol. 40, #1, January 1998;
   5 Vol. 40, #3, March 1998;
   6 Vol. 40, #5, May 1998;
   7 Vol. 40, #9, September 1998;
   8 Vol. 40, #12, December 1998;
   9 Vol. 41, #5, May 1999;
   10 Vol. 42, #4, April 2000;
   11 Vol. 42, #10, October 2000;
   12 Vol. 43, #1, January 2001;
   13 Vol. 43, #4, April 2001 (referred to as Varfolomeyev, 1995–2001).

[5] Harford, Jim: *Korolev – how one man masterminded the Soviet drive to beat America to the moon*. John Wiley & Sons, New York, 1997.

[6] Khrushchev, Sergei: The first Earth satellite – a retrospective view from the future. From: Roger Launius, John Logsdon and Robert Smith: *Reconsidering Sputnik – 40 years since the Soviet satellite*. Harwood Academic, Amsterdam, 2000.

[7] Matson, Wayne R: *Cosmonautics – a colourful history*. Cosmos Books, Washington DC, 1994.

[8] Siddiqi, Asif: *The challenge to Apollo*. NASA, Washington DC, 2000.

2

The first moon probes

Sputnik changed everything. Most of the great historical events of our time make an immediate impact that fades over time. Sputnik was different. When the first Earth satellite was launched, Soviet leader Nikita Khrushchev calmly took the call from Baikonour Cosmodrome, thanked Korolev courteously and went to bed. *Pravda* did report the launching the next day, but well down the page, blandly headed 'Tass communiqué'. In the West, the British Broadcasting Corporation announced the launching at the end of its late news bulletin, a certain vocal hesitancy indicating that neither the station nor the announcer knew exactly what to make of this strange event.

Whatever the political leadership, ordinary people knew. As Korolev and his colleagues took the long train journey back from Baikonour Cosmodrome, people came onto the platforms to stop the train and meet the engineers concerned. There was a palpable, rising air of excitement as they drew close to Moscow. By this stage, people throughout the Soviet Union were talking and chattering about this extraordinary event. The next day, *Pravda* made the satellite the *only* front-page story. Khrushchev was soon bragging about its achievements to foreign leaders.

But this was as nothing compared with the pandemonium in the United States, which ranged between the hysterical and apoplectic. There was admiration for the Soviet achievement, but it was couched in the regret that America had not been first. The Americans had been publicizing their plans to launch an Earth satellite for a number of years. Because America was the world's leading technological country, nobody had ever suggested that the United States might not be first: the very idea was unthinkable. In fact, the Russians had also been broadcasting, very publicly and in some technical detail, their intentions of launching an Earth satellite. Although the first Sputnik was a surprise to the American people, it was not a surprise to the Soviet people, who had been expecting it and for whom space travel had achieved an acceptance within popular culture from the 1920s, renewed in the 1950s.

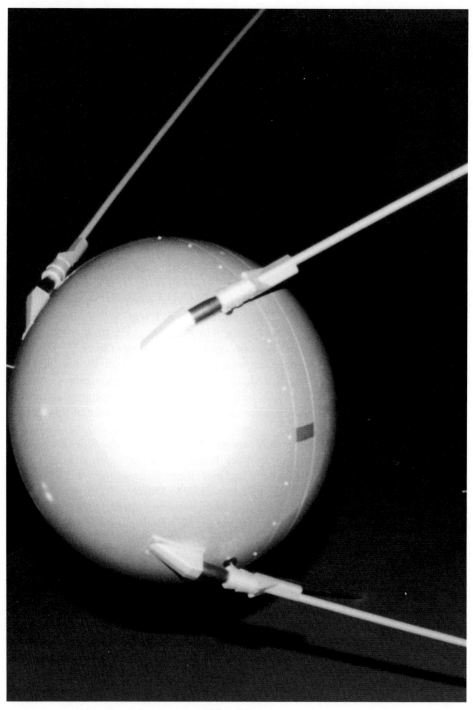

Sputnik

The first Sputnik made a deep impact on ordinary Americans. The rocket body of the Sputnik entered orbit and could be seen tracking across the cooling night autumn skies of North America. The long aerials on the back of the spacecraft transmitted *beep! beep! beep!* signals that could be picked up by relatively simple receivers. Within months, polling found that almost all Americans had heard of the Sputnik and nearly everyone had comments to make about it.

1958 PROPOSALS FOR MOON ROCKETS

Taking advantage of the public reaction, Korolev set new tasks for Mikhail Tikhon-ravov and his 'planning department for the development of space apparatus'. Korolev had first raised the idea of lunar exploration with government as far back as a meeting on 30th April 1955, but nothing had come of it. Now the climate was quite different. Popular enthusiasm for spaceflight had reignited in a surge of public interest reminiscent of the glorious 1920s and early 1930s. Three design groups were set up: one for a manned spacecraft, one for communications satellites and one for automatic lunar spacecraft. The lunar group was put under the charge of a brilliant young designer, then only 30 years old, Gleb Yuri Maksimov (1926–2000).

Based on discussions with him, on 28th January 1958, Tikhonravov and Korolev sent a letter to the Central Committee of the Communist Party of the Soviet Union and the government called *On the launches of rockets to the moon* (sometimes also translated as *A programme for the investigation of the moon*). This proposed that two

Gleb Yuri Maksimov

spacecraft be sent to the moon. One would hit the moon, while the other would take photographs of its hidden far side and transmit them to the Earth by television. The impacting probe would signify its arrival either by the cessation of its telemetry signal, or through the igniting of explosives, which could be seen from the Earth. The government agreed to the proposal within two months, on 20th March 1958. Contrary to Western impressions that Soviet spaceshots were ordered up by the political leadership, the opposite is true. Most of the early Soviet space missions resulted from proposals by the engineers, convincing government of the political and publicity advantages.

Tikhonravov and Korolev followed this with a grander plan for space exploration that summer. Called *Most promising works in the development of outer space*, this was an audacious plan outlining a vast programme of space exploration – variations of this title have also appeared (e.g., *Preliminary considerations for the prospects of the mastery of outer space*), probably a function of translation. None of this was evident in the West – indeed, details of *Most promising works in the development of outer space* were not published until decades later. Yet, the plan outlined, at this extraordinarily early stage, how the Soviet Union was to conquer the cosmos. This was what they proposed:

- Small research station of 15 to 20 kg to land on the moon.
- Satellite to photograph the lunar surface.
- Upgrade the R-7 launcher to four stages to send a probe to orbit the moon and return films to Earth.
- Send robotic spacecraft to Mars and Venus.
- Develop Earth orbit rendezvous.
- Manned spacecraft for flight around the moon.
- Eventually, manned flights to the moon, Venus and Mars.
- Goal of permanent colony on the moon.
- Development of the critical path technologies for rendezvous, life support systems and long-distance communications.

Now Maksimov's group soon came up with its first set of detailed designs. Four types of spacecraft were proposed. They were called the *Ye* or *E* series, after the sixth letter in the Russian alphabet (the first five had already been assigned to other projects). These are shown in Table 2.1.

These plans were soon modified. The Ye-4 probe was the problem one. Nuclear experts warned that a nuclear explosion on the moon would, without an atmosphere there, be difficult to observe and that the visibility of even a conventional explosion was uncertain, so this probe was dropped. The engineers also worried about how to track a small spacecraft *en route* to the moon. They came up with the idea of fitting 1 kg of sodium or barium to be released during the journey. This would create a cloud of particles that could be spotted by the right sensors. Once the moon had been hit (Ye-1), they would move quickly on to farside photography missions (Ye-2, 3).

Table 2.1. Plans for first generation of Soviet moon probes, OKB-1, spring 1958.

Name	Weight (kg)	Objective	Notes
Ye-1	170	Lunar impact	Five scientific instruments
Ye-2	280	Farside photography	Six scientific instruments Two lenses: 200 mm and 500 mm
Ye-3	280	More detailed farside photography	Camera lens of 750 mm
Ye-4	400	Lunar impact	Carrying nuclear or conventional explosives

AN UPPER STAGE: ENTER SEMYON KOSBERG

Spacecraft design was only one part of the jigsaw required to put the moon project together. The other crucial part was an upper stage able to send the probe toward the moon. The rocket that had launched Sputnik, Sputnik 2 and 3 – the R-7 – was capable of sending only 1,400 kg into low-Earth orbit, no further. A new upper stage would be required. Back in April 1957, Mikhail Tikhonravov had suggested that it would be possible to send small payloads to the moon, through the addition of a small upper stage to the R-7.

Chronology of the early Soviet lunar programme
4 Oct 1957   Sputnik.
28 Jan 1958   Proposal to government by Korolev and Keldysh.
10 Feb 1958   Agreement with OKB-154 (Kosberg) for upper stage.
20 Mar 1958   Approval by government of proposal for moon probe.
5 Jul 1958   *Most promising works in the development of outer space.*

Korolev considered two options for an upper stage. First, he turned to the main designer of rocket engines in the Soviet Union, Valentin Glushko. Glushko had designed the main engines for the R-7, the kerosene-propelled RD-107 and RD-108 (RD, or rocket engine, in Russian *Raketa Digvatel*). Since then, though, he had discovered UDMH, or to be more correct, it had been discovered by the State Institute for Applied Chemistry. UDMH stood for unsymmetrical dimethyl methyl hydrazine and it had many advantages. When mixed with nitric acid or one of its derivatives, this produced powerful thrust for a rocket engine. Unlike liquid oxygen – which must be cooled to very low temperatures – and kerosene, UDMH and nitric acid could be kept in rockets and their adjacent fuelling tanks at room temperature for some time and for this reason were called 'storable' fuels.

They were hypergolic and fired on contact with one other, saving on ignition systems. The great disadvantage was that they were toxic: men working on them had to wear full proper protective gear. The consequences of an unplanned explosion did not bear thinking about and Korolev labelled the fuel 'the devil's own venom'. Glushko proposed the R-7 fly his new upper stage, the RD-109.

Korolev had his doubts as to whether Glushko could get his new engine ready for him in any reasonable time. He learned that an aircraft design bureau, the OKB-154 of Semyon Kosberg in Voronezh, had done some development work on a restartable rocket engine using the tried-and-tested liquid oxygen and kerosene. Semyon Kosberg was not a spacecraft designer: his background was in the Moscow Aviation Institute, he built fighters for the Red Air Force and his interest was in aviation. Korolev, wary of Glushko's engine and skeptical of his ability to deliver on time, persuaded Kosberg to build him a small upper stage and they signed an agreement on 10th February 1958, even before government agreement for the moon programme. The new engine, later called the RD-105 (also referred to as the RD-0105 and the RO-5), was duly delivered only six months later, in August 1958. It was the first rocket designed only to work in a vacuum. This new variant of the R-7 was given the technical designation of the 8K72E (a more powerful version of the upper stage later became the basis of the first manned spaceship, Vostok, and was known as the 8K72K).

R-7 rocket, with upper stage block E for lunar missions (8K72E)

Length	33.5 m
Diameter (blocks ABVGD)	10.3 m
Weight	279.1 tonnes
of which frame	26.9 tonnes
propellant	256.2 tonnes
Thrust at liftoff	407.5 tonnes

8K72E upper stage (block E)

Length	5.18 m
Diameter	2.66 m
Weight	8,510 kg
Frame	1,110 kg
Propellants	6,930 kg

RD-105 engine

Weight	125 kg
Thrust	5.04 tonnes
Fuel	LOX and kerosene
Pressure	46 atmospheres
Specific impulse	316

Burn times

Burn time block A	320 sec
Burn times blocks BVGD	120 sec
Burn time block E	790 sec

Source: Varfolomeyev (1995–2001)

A suborbital flight of the new moon rocket took place on 10th July 1958. The aim was to test the control system for the ignition and separation of the upper stage, but the mission never got that far, for the rocket blew up a few seconds after liftoff.

A TRACKING NETWORK

The moon programme required a tracking network. To follow Sputnik, a government resolution had been issued on 3rd September 1956 and authorized the establishment of up to 25 stations [1]. By the time of Sputnik, about 13 had been constructed, the principal ones being in Kolpashevo, Tbilisi, Ulan Ude, Ussurisk and Petropavlovsk, supplemented by visual observatories in the Crimea, Caucasus and Leningrad.

For the moon programme, systems were required to follow spacecraft over half a million kilometres away. For this, a new ground station was constructed and it was declared operational on 23rd September 1958, just in time for the first Soviet lunar probe. Yevgeni Boguslavsky, deputy chief designer of the Scientific Research Institute of Radio Instrument Building, NII-885, was responsible for setting up the ground station. It was located in Simeiz, at Kochka Mountain in the Crimea close to the Crimean Astrophysical Observatory of the Physical Institute of the USSR Academy of Sciences. His choice of the Crimea was a fateful one, for all the main subsequent Soviet observing stations came to be based around there, including the more substantial subsequent interplanetary communications network. Boguslavsky obtained the services of military unit #32103 for the construction work and it was sited on a hill facing southward onto the Black Sea. Sixteen helice aerials were installed, turning on a cement tower. A backup station was also built in Kamchatka on the Pacific coast.

Although the station was declared operational, the people working there might have taken a different view, for the ground equipment was located in trailers, ground control was in a wooden barrack hut, many of the staff lived in tents and food was supplied by mobile kitchen. All of this cannot have been very comfortable in a Crimean winter.

The Soviet Union also relied on a 24 m parabolic dish radio telescope in Moscow and the receiver network used for the first three Sputniks. Pictures of the first missions – which indicated a location 'near Moscow' – showed technicians operating banks of wall computers and receiving equipment, using headphones, tuners and old-fashioned spool tape recorders, printing out copious quantities of telex. Presumably, they didn't wish to draw the attention of the Americans to their new facilities on the Black Sea and this remained the case until 1961, by which time it was guessed, correctly, that the Americans had found out anyway.

Early tracking dish, Crimea

ONLY HOURS APART: THE MOON RACE, AUTUMN 1958

By this time, the United States had launched their first satellite (Explorer 1, January 1958) and had made rapid progress in preparing a lunar programme. Korolev followed closely the early preparations by the United States to launch their first moon probe, called Pioneer. Learning that Pioneer was set for take-off on 17th August 1958, Korolev managed to get his first lunar bound R-7, with its brand-new Kosberg upper stage, out to the pad the same day, fitted with a Ye-1 probe to hit the lunar surface. The closeness of these events set a pattern that was to thread in and out of the moon programmes of the two space superpowers for the next eleven years.

There had been a lot of delays in getting the rocket ready and Korolev only managed to get this far by working around the clock. The lunar trajectory mapped out by Korolev and Tikhonravov was shorter than Pioneer. Korolev waited to see if Pioneer was successfully launched. If it was, then Korolev would launch and could still beat the Americans to the moon. Fortunately for Korolev, though not for the Americans, Pioneer exploded at 77 sec and a relieved Korolev was able to bring his rocket back to the shed for more careful testing.

A month later, all was eventually ready. The first Soviet moon probe lifted off from Baikonour on 23rd September 1958. Korolev may have worried most about whether the upper stage would work or not, but the main rocket never got that far, for vibration in the BVGD boosters caused it to explode after 93 sec. Despite launching three Sputniks into orbit, the R-7 was still taking some time to tame. Challenged about

Sergei Korolev at launch site

repeated failures and asked for a guarantee they would not happen again, Korolev lost his temper and yelled: *Do you think only American rockets explode?*

The August drama came around a second time the following month. At Cape Canaveral, the Americans counted down for a new Pioneer, with the launch set for 11th October. In complete contrast to the developments at Cape Canaveral, which were carried out amidst excited media publicity, not a word of what was going on in Baikonour reached the outside world. Again, Korolev planned to launch the Ye-1 spaceship on a faster, quicker trajectory after Pioneer. News of the Pioneer launching was relayed immediately to Baikonour, Korolev passing it on in turn over the loudspeaker.

Not long afterwards, the news came through that the Pioneer's third stage had failed. Korolev and his engineers now had the opportunity to eclipse the Americans. On 12th October, his second launching took place. It did only marginally better than the previous month's launch, but the vibration problem recurred, blowing the rocket apart after 104 sec. Although Pioneer 1 was launched thirteen hours before the Soviet moon probe was due to go, the Russian ship had a shorter flight time and would have overtaken Pioneer at the very end. Korolev's probe would have reached the moon a mere six hours ahead of Pioneer. According to Swedish space scientist and tracker Sven Grahn who calculated the trajectories many years later, 'the moon race never got much hotter!'.

These two failures left Korolev and his team downcast. Although the R-7 had given trouble before, two failures in a row should not be expected, even at this stage of its development. Boris Petrov of the Soviet Academy of Sciences was appointed to head up a committee of inquiry while the debris from the two failures was collected and carefully sifted for clues. What they found surprised them. It turned out that the Kosberg's new upper stage, even though it had never fired, was indirectly to blame. The new stage, small though it might be, had created vibrations in the lower stage of the rocket at a frequency that had caused them to break up. This was the first, but far from the last, time that modification to the upper stages of rockets led to unexpected consequences.

Devices were fitted to dampen out the vibration. Although they indeed fixed this problem, the programme was then hit by another one. It took two months, working around the clock, to get a third rocket and spacecraft ready. The third rocket took off for the moon on 4th December. As it flew through the hazardous 90–100 sec stage, hopes began to rise. They did not last, for at 245 sec, the thrust fell to 70% on the core stage (block A) and then cut out altogether. The rocket broke up and the remnants crashed downrange. The crash was due to the failure of a hydrogen peroxide pump gearbox, in turn due to the breaking of a hermetic seal which exposed the pump to a vacuum. It must have been little consolation to Korolev that the next American attempt, on 6th January, was also a failure, though it reached a much higher altitude, 102,000 km.

The Soviet failures were unknown except to those directly involved and the political leadership. America had experienced its own share of problems, but there the mood was upbeat. The probes had a morale-boosting effect on American public opinion. There was huge press coverage. The Cape Canaveral range (all it had been to date was an air force and coastguard station) became part of the American consciousness. Boosters, rockets, countdowns, the moon, missions, these words all entered the vocabulary. America was fighting back, and if the missions failed, there were credits for trying.

On the Russian side, there was little public indication that a moon programme was even under way. In one of the few, on 21st July 1957, Y.S. Khlebstsevich wrote a speculative piece outlining how, sometime in the next five to ten tears, the Soviet Union would send a mobile caterpillar laboratory or tankette to rove the lunar surface and help choose the best place for a manned landing [2]. Information about the Soviet space programme, which had been relatively open about its intentions in the mid-1950s, now became ever more tightly regulated. Chief ideologist Mikhail Suslov laid down the rubric that there could not be failures in the Soviet space programme. Only successful launchings and successful mission outcomes would be announced, he decreed, despite the protests at the time and later of Mstislav Keldysh. A cloud of secrecy and anonymity descended. The names of Glushko and Korolev now disappeared from the record, although they were allowed to write for the press under pseudonyms. Sergei Korolev became 'Professor Sergeev'. Valentin Petrovich Glushko's *nom de plume* was only slightly less transparent: 'Professor G.V. Petrovich', for it used both his initials (in reverse) and his patronymic.

So whenever spaceflights went wrong, their missions were redefined to prove that they had, indeed, achieved all the tasks set for them. This was to lead Soviet news management, in the course of lunar exploration, into a series of contradictions, blunders, disinformation, misinformation and confusion. But it was best, as in the case of the first three moonshots, that nothing be known about them at all.

FIRST COSMIC SHIP

Undeterred though undoubtedly disappointed, Korolev hoped to be fourth time lucky. He aimed to make his fourth attempt for New Year's Day. Preparing the rocket in such record times was extremely difficult and the engineers complained of exhaustion. Baikonour was now in the depths of winter and temperatures had fallen to $-30°$C. There were two days of delays and the probe was not launched until the evening of 2nd January 1959.

Blocks B, V, G and D fell away at the appropriate moment. The core stage, the block A, cruised on. The time came for block A to fall away. Now, Semyon Kosberg's 1,472 kg small upper stage faced its crucial test. With apparently effortless ease, the stage achieved escape velocity (40,234 km/hour) and headed straight moonwards. The final payload, including the canister, sent moonbound weighed 361 kg, but the actual moon probe was 156 kg. The spacecraft was spherical and although the same shape as the first Sputnik was four times heavier, with a diameter of 80 cm, compared with the 56 cm of Sputnik. It was pressurized and the four antennae and scientific instruments popped out of the top. Signals would be sent back to Earth on 183.6 MHz for trajectory data and 19.993 MHz for scientific instruments (this is called 'downlink') and commands sent up on 115 Hz ('uplink'). The radio system had been designed and built by Mikhail Ryanzansky of the NII-885 bureau, one of the original Council of Designers. To save battery, signals would be sent back for several minutes or longer at a time at pre-timed intervals, but not continuously. The upper stage also had a transmitter which sent back signals in short bursts every 10 sec for several hours as it headed into deep space.

The spacecraft carried instruments for measuring radiation, magnetic fields and meteorites. The magnetometer was only the second carried by a Soviet spaceship and

First Cosmic Ship launch

First Cosmic Ship

arose from a 1956 meeting between chief designer Sergei Korolev and the first head of the space Magnetic Research Laboratory, Shmaia Dolginov (1917–2001) [3]. He headed the laboratory in the Institute of Terrestrial Magnetism (IZMIRAN) where he had mapped the Earth's magnetic field by sailing around the world in wooden ships using no metallic, magnetic parts. He worked with Korolev to install a magnetometer on Sputnik 3, which duly mapped parts of the Earth's magnetic field. Now they would be installed on lunar probes to detect magnetic fields around the moon. The magnetometer was called a triaxial fluxgate magnetometer with three sub-instruments and sensors with a range of −3,000 to 3,000 gammas.

Similarly, ion traps first flown on Sputnik 3 would be used on the lunar probe. Ion traps were used to detect and measure solar wind and solar plasma and were developed by Konstantin Gringauz (1918–1993), who had been flying his traps on sounding rockets as far back as the 1940s. He had famously built the transmitter on Sputnik and was the last man to hold it before it was put in its carrier rocket. The meteoroid detector was developed by Tatiana Nazarova of the Vernadsky Institute. Essentially, it comprised a metal plate on springs which recorded any impact, however tiny. The cosmic ray detector was developed by Sergei Vernov (1910–1982) of the Institute of Nuclear Physics in Moscow, who had been flying cosmic ray detectors on balloons since the 1930s.

Instruments on the First, Second Cosmic Ship
Gas component detector.
Magnetometer (fields of Earth and moon).
Meteoroid detector.
Cosmic ray detector.
Ion trap.
1 kg of sodium vapour.

As the probe moved rapidly between 20,000 km and 30,000 km out from Earth, it was possible to use the radio signals to make very precise measurements of its direction and velocity. From these, it was apparent that the spacecraft would not hit the moon after all, though unlike the American spacecraft it would not fall back to Earth. On 3rd January, when 113,000 km out from Earth, the spacecraft released a golden-orange cloud of sodium gas so that astronomers could track it. The cloud was visible in the sky over the Indian Ocean and it confirmed that the probe would come quite close to the moon.

One problem was: what to call it? In Moscow, it was referred to as 'The First Cosmic Ship' because it was the first spacecraft to leave the Earth's gravitational sphere of influence at escape velocity. The Russians appeared reluctant to name it a moon probe, because that would imply that it was supposed to impact on the moon, which of course it was. Already, the Suslov decision was having its baleful impact. On 6th January, Anatoli Blagonravov of the Academy of Sciences denied flatly that it was ever intended to hit the moon but to pass close by instead [4]. Later, in 1963, it was retrospectively given the name of Luna 1. In the West, the first three probes were called Lunik, but this was a media-contrived abbreviation of 'Luna' and 'Sputnik' and was never used by the Russians themselves. Several of the early designators for the Soviet space programme were unclear and applied inconsistently, but thankfully never as confusingly so as the early Chinese space programme.

On 4th January, the First Cosmic Ship passed by the moon at a distance of 5,965 km some 34 hours after leaving the ground. It went on into orbit around the Sun between the Earth and Mars between 146.4 million kilometres and 197.2 million kilometres. The probe was a dramatic start to moon exploration: it ventured into areas of space never visited before. Signals were picked up for 62 hours, after which the battery presumably gave out, at which point the probe was 600,000 km away.

The first round of results was published by scientists Sergei Vernov and Alexander Chudakov in *Pravda* on 6th March 1959. More details were given by the president of the Academy of Sciences, Alexander Nesmyanov, opening the Academy's annual general meeting that spring, which ran from 26th to 28th March. First, no magnetic field was detected near the moon, but scientists were aware that it was possibly too far out to detect one. The magnetometer noted fluctuations in the Earth's magnetic field as the First Cosmic Ship accelerated away. A contour map of the Earth's radiation belts was published, showing them peak at 24,000 km and then fall away to a low level some 50,000 km out. Second, the meteoroid detector, which was calibrated to detect dust of a billionth of a gramme, suggested that the chances of being hit by dust on the way out to the moon or back was minimal. Third, in a big finding, Konstantin

First Cosmic Ship, top stage

Gringauz's ion traps detected how the Sun emitted strong flows of ionized plasma. This flow of particles was weak, about 2 particles/cm^2/sec, because the sun was at the low point in its cycle, but the ship's ion traps had determined the existence of a 'solar wind'. This was one of the discoveries of the space age and Gringauz estimated that the wind blew at 400 km/sec [5].

THE BIG RED LIE?

The First Cosmic Ship was denounced in some quarters of the Western world as a fraud and one writer, Lloyd Malan, even wrote a book about it called *The big red lie*.

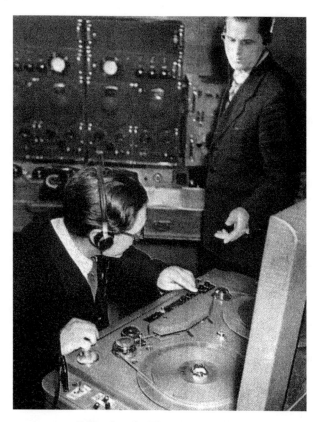

Following the First Cosmic Ship

The reason? Few people in the West picked up its signals, even though the Russians had, as usual, announced their transmission frequencies (183.6, 19.993 and 19.997 MHz). Not only that, but the original Tass communiqué announcing the mission had told observers when the moonship would be over Hawaii, when the sodium cloud would be released and even where to look for it (the constellation Virgo).

The explanations were mundane, rather than conspiratorial. The Russians had inconveniently launched the First Cosmic Ship late on a Friday night and most professional observers had long since gone home for the weekend. By the time the Earth had rotated in line of sight for American observatories, the First Cosmic Ship was already well on its way and ever more difficult to pick up. In the event, signals were received on the next day by Stanford University in California when it was about 171,000 km out. At the Jet Propulsion Laboratory in California, staff were recalled over the weekend in a frantic effort to locate the spacecraft, which they eventually did when it was 450,000 km out, eight hours after it passed the moon. American military signal stations probably also tracked the spacecraft in Hawaii, Singapore, Massachusetts and Cape Canaveral, but if they received signals, they never told.

In Britain, the director of the large radio telescope at Jodrell Bank, Bernard Lovell, was at home listening to Johann Sebastian Bach's *Fantasy and fuge*. Jodrell Bank had been established by a physics professor, Bernard Lovell, who had spent the war developing radar to detect enemy planes and ships. In peacetime, he now adapted ex-army radars to study cosmic rays and meteor trails. This work was so promising that in 1950 he got the go-ahead for a large radio telescope for radio mapping of deep space objects and this was, fortuitously, completed just in time for the launching of Sputnik seven years later. There was some debate in Jodrell Bank as to whether the huge dish telescope should be used to track spacecraft at all, but the station had considerable financial liabilities and the glow of world media publicity attached to the station's role in tracking spacecraft soon enabled that debt to be cleared. In fact, it was not the Russians but the Americans who first brought Jodrell Bank into the moon programme, paying for the use of its facilities in 1958 for the early American moon probes. Jodrell Bank had tried but failed to pick up the First Cosmic Ship, but, Bernard Lovell added, the station still believed that the probe existed! He put down his failure to obtain signals as due to inexperience. He had imagined that it would transmit continuously and had not understood the Russian system of periodic transmission, the 'communications session' [6].

The early moon shots of the United States and the Soviet Union had much in common. The first and the most obvious was their high failure rate. With the successful launching of the First Cosmic Ship, Russia and America had each tried four times. One Russian probe had reached but missed the moon. One American probe had reached 113,000 km, the other 102,000 km before falling back. All the rest had exploded early on.

Here, the similarities ended. The Russian Ye-1 probe was large, weighing 156 kg, with a simple (albeit elusive) objective: to impact on the moon. Six instruments were carried. By contrast, the American Pioneer probes were tiny, between 6 kg and 39 kg. They carried similar instruments: for example, like the early Russian probes, Pioneer 1 carried a magnetometer. The early American missions were more ambitious, aiming for lunar orbit and to take pictures of the surface of the moon. The camera system on Pioneer was tiny, weighing only 400 g, comprising a mirror and an infrared thermal radiation imaging device.

The First Cosmic Ship was hailed as a great triumph in the Soviet Union. The third year of space exploration could not have opened more brightly. Stamps were issued showing the rocket and its ball-shaped cargo curving away into a distant cosmos.

SECOND COSMIC SHIP

Although there was much celebration at the achievement of the First Cosmic Ship, Korolev still faced the task of hitting the moon and doing so before the Americans. In March, the Americans at last passed the moon, but the accuracy of Pioneer 4 was much less than the First Cosmic Ship, for Pioneer 4 missed the moon by 60,015 km. The first half of 1959 saw continued Soviet difficulties with the R-7 launcher and a new one was not ready until the summer. The moon probe itself was slightly modified, the payload being heavier at 390 kg and received the designator Ye-1a. The first attempt at launch had been planned for 16th June, but the upper stage had been incorrectly fuelled and had to be unloaded and then refilled. It did not matter in the end, for when the probe was launched two days later on 18th June 1959, the inertial guidance system failed at 152 sec and the rocket crashed out of control and was exploded on ground command.

The fix took three months and the next rocket counted down on 9th September. Ignition took place, but the engines did not build up sufficient thrust for the rocket to take off. This was what became known in American terminology as a pad abort. Korolev must have been extremely frustrated at this stage, for 13 months after the first attempt, he still had not hit the moon.

The 390.2 kg Second Cosmic Ship was eventually sent up on 12th September. The upper stage reached the intended escape velocity of 11.2 km/sec and then the spacecraft separated from the upper stage. Transmissions began at once, using three transmitters working on 183.6, 19.993 and 39.986 megacycles. The signals told ground control that its course was, this time, dead centre and Radio Moscow quickly announced that the rocket would reach the moon at 00:05 on the 14th September. The ship spun slowly around its axis, once every 14 min. The final stage also sent back radio signals on 20 MHz and 19 MHz as it headed away.

To mark the visual progress of the rocket, the Second Cosmic Ship released a sodium vapour cloud on the 13th, some 156,000 km out. It eventually expanded into a 650 km diameter cloud and this was spotted by observatories in Alma Ata, Byurakan, Abastuma, Tbilisi and Stalinabad. The Second Cosmic Ship carried an identical suite of scientific instruments to the first, although Shmaia Dolginov's magnetometer had been modified to reduce the range of measurement to between −750 and +750

The Second Cosmic Ship

gammas, where a response was considered more likely. It took measurements every minute during the flight out and confirmed the observations of the First Cosmic Ship. As the probe neared the moon, the instruments were working perfectly and were searching for lunar magnetic and radiation fields (none were found when the last measurement came in 55 km out). The Second Cosmic Ship not only encountered the solar wind met by its predecessor, but measured it. Other instruments measured alpha particles (nuclei of carbon, nitrogen), X-rays, gamma rays, high- and low-energy electrons and high-energy particles.

Korolev and the designers gathered in the control room as the Second Cosmic Ship neared the moon. The Russians had been stung by the claims that the First

Sodium release, the Second Cosmic Ship

Cosmic Ship had been a fraud, or 'a big red lie' and this time took no chances. At Jodrell Bank, the Russians had again inconveniently launched a moon probe during a weekend. Bernard Lovell captained his local cricket team and refused all remonstrations to be interrupted to track the new spaceship. He was eventually persuaded to return to the observatory where a telex, hot in from Moscow, gave him not only the transmission and trajectory details but the intended time of impact. Cricket or not, this was a serious world event now. Jodrell Bank started tracking the Second Cosmic Ship the moment the moon rose over the horizon some five hours before impact was due. Round the world, radio stations went on a night-time vigil to wait for the historic moment of impact, in what must have been the first worldwide news coverage of an event taking place away from the Earth. Signals poured back loud and clear from the spacecraft against the eternal static of deep space. The Second Cosmic Ship plunged into the moon's gravity well at an angle of $60°$ and a velocity of almost 3 km/sec. Then in a instant, 2 min and 24 sec after midnight, the signals were abruptly cut short and there was dead silence!

The Second Cosmic Ship had made it, reached the moon and impacted onto it at great speed. It was a bull's eye, barely $1°$ west longitude and $30°$ north latitude. The Second Cosmic Ship crashed somewhere in a triangle shaped by the craters Archimedes, Aristillus and Autolycus in the small *mare* (sea) called the Marsh of Decay (*Palus Putredinis*), scattering hammer-and-sickle ball-shaped momentos onto the lunar surface to mark the occasion. The upper stage of the rocket followed 30 min later, though it carried no transmitter so its impact point is unknown. And the person who told the waiting world was Bernard Lovell, who got the news out first, for he confirmed that the signals had ceased and that the trajectory had intercepted the moon. Still some Americans denied the Soviet achievement. Bernard Lovell calculated the Doppler shift on the signals, proving that they came from a moving object falling fast toward the centre of the moon. He played the tape recording, with the abrupt stop, over the phone to the New York media and that seemed to satisfy most of them. By way of a thank-you for the telex, many years later he handed a tape recording of the signals to Mstislav Keldysh.

No one was more pleased than Nikita Khrushchev. He was able straightaway to present President Dwight Eisenhower with a model of the commemorative pennants which his country had just deposited on the moon. Khrushchev loved these gestures. Not only had the USSR reached the moon, but he could bring the good news in the latest Soviet aircraft. First, there was the Tupolev 104, the first successful modern jetliner. Then the Soviet Union developed the Tupolev 114. This was a massive, fast, long-range propellor-driven airliner able to fly 220 people with two decks high above the clouds. Khrushchev amazed the Americans when he flew to New York in this huge silver plane without ever stopping once for refuelling.

The scientific results of the mission of the Second Cosmic Ship were published the following spring. To do so, scientists went through 14 km of teletype! Neither a magnetic field nor a radiation belt was found around the moon. The outer belt of electrons in the Earth's charged particles reached out as far as 50,000 km. The four ion traps on the outside measured the flows of the currents of ion particles all the way out to the moon. Their concentration varied, sometimes as less than 100 particles/cm^3.

Sergei Vernov

But 8,000 km out from the moon, current intensities increased, suggesting the existence of a shell of a lunar ionosphere.

THE AUTOMATIC INTERPLANETARY STATION

The early frustrations of 1958 could be put to one side now: the knifelike precision of the Second Cosmic Ship showed what could be done. The accuracy of the Second Cosmic Ship was not lost on the Americans, who had never attained such early accuracy. Not that they were given much time to recover. Three weeks later, and on the second anniversary of Sputnik, a third cosmic ship lifted off the pad.

This was the first Ye-2 type of moon probe (though, to be completely accurate, it had now been designated Ye-2a). There was quite a jump between the Ye-1 type of probe and the ambition of a Ye-2 or Ye-3. Both required great accuracy, but the farside photography mission especially so. For lunar imaging, Keldysh's Mathematical Institute was called in. Such a mission must take place when the farside was lit up by the sun and bring the probe on a trajectory back to the Earth high over the Soviet Union so that it could transmit back the pictures. Such optimum conditions would take place infrequently: in October 1959 (photography after approaching the moon) and April 1960 (photography while approaching the moon). The spacecraft would require an orientation system to make sure the cameras pointed the right way and that the transmissions were subsequently relayed back to the Earth. The orientation system was developed by Boris Raushenbakh and a team of seven young engineers who built the parts from shop-bought electronic components, the Soviet *Radio Shack* of its day. Boris Raushenbakh (1915–2001) was, as his name suggests, German by background and for this reason was interned during the war. In his spare time, he developed an expertise in the history of Russian art. He was allowed to return to the Keldysh Research Centre after the war, where he developed a knowledge of spacecraft orientation. Gas jets provided the all-important orientation system. Sensors were used to maintain orientation toward the Earth, sun and moon. The station was the first spacecraft to develop a three-axis stabilization system. For

The Automatic Interplanetary Station

the flyby, the sensors would be used to locate the sun, Earth and moon and once this was done, the spacecraft's thrusters would fire until it was brought into the desired position for photography or communications or whatever was required. His system has been used ever since.

Two camera systems were developed, the successful one being built by Television Scientific Research Institute NII-380 in Leningrad under Petr Bratslavets (1925–1999), assisted by I.A. Rosselevich. To take pictures, the Russians opted not for relatively new television systems like the Americans but for older, mechanical designs likely to give much higher quality. The imaging system was called Yenisey 2 and this comprised a duel-lens camera, scanner and processing unit. The dual lens could take up to 40 pictures at 200 mm, f5.6 (designed for the full moon) or 500 mm, f9.5, designed for close-ups. The cameras could not be moved or swivelled: instead, the spacecraft itself would be rotated to point in the appropriate direction. Transmissions of signals could be made at two speeds: slow, at 1.25 lines a second (for distant transmissions) and faster, at 50 lines a second (closer to Earth).

The photographs would be developed onboard and then scanned by a television camera. This system was designed by Scientific Research Institute for Radio Instrument Building, NII-885, where the person responsible was the deputy chief designer Yevgeni Boguslavsky (1917–1969) who used, instead of the traditional valves, some of the new transistors. The station was the first to make use of transistors. Now long outdated, transistors were new in the 1950s, the first being made by the NPO Svetlana in Leningrad in 1955. The first transistors had been flown in Sputnik 3 the previous year, but this was the first time that they were the basis of the electrical system. Boguslavsky had developed optical and radio tracking systems for missiles in the 1940s and had been involved in the radio tracking of the First and Second Cosmic Ships. As the probes swung back to Earth, the television camera would scan the photographs and transmit them by radio. Transmission would be by omnidirectional antenna, sending signals out over a broad range, which improved the chances of them being picked up but diminished the quality of the signal received. Transmissions were to be sent on two frequencies: 39.986 MHz and 183.6 MHz, using a system of impulse transmitters able to achieve high rates of telemetry. The Ye-2 was probably the most complex spacecraft in the very early days of space exploration. The Ye-3 was an even more sophisticated system, but was cancelled when it was decided to concentrate on the Ye-2 versions, the Ye-2a and Ye-2f.

Years later, it emerged that the Soviet specialists had not been able to manufacture radiation-hardened film that would survive the journey through the radiation belts and the translunar environment. Instead, they used American film retrieved from Gentrix balloons – spy balloons floated across the Soviet Union from American bases in western Europe to spy on military facilities but whose film was known to be radiation-protected.

The weight of the new lunar craft was 278 kg. The Ye-2 looked quite different from the Ye-1, being a cylindrical canister with solar cells of the type already used on Sputnik 3. The Ye-2 was 1.3 m tall, 1.2 m in diameter at the widest but 95 cm for most of its body. The cannister was sealed and pressurized at 0.23 atmospheres. Shutters opened and closed to regulate the temperature, being set to open if it rose above 25°C. Four antennae poked out through the top of the spacecraft, two more from the bottom. The cameras were set in the top and the other scientific instruments were mounted on other parts of the outside. In addition to the cameras, the main payload, the spacecraft carried a cosmic ray detector and micrometeoroid detector.

The new moon probe arrived at Baikonour in August 1959, even before the mission of the Second Cosmic Ship. There was still some testing to be completed there and this was signed off on 25th September. Launching took place on 4th October, two years after Sputnik. The new launching caused mystery at first. Far from taking a rapid course out to the moon, it swung lazily outward in what was actually an irregular high-Earth orbit, 48,280 km by 468,300 km, inclination 55°. The trajectory had been carefully calculated with the help of a computer at the Department of Applied Mathematics of the Steklov Institute of the USSR Academy of Sciences. This time it was curiously labelled the 'Automatic Interplanetary Station' (AIS). The Russians announced its transmission frequencies 39.986 MHz (science) and 183.6 MHz (trajectory). They informed Jodrell Bank, which picked up the station some ten hours after launch. The Jodrell Bank staff had to do this without their director. Bernard Lovell was on a visit to the United States. His NASA hosts were giving him a mock journey to the moon in newly opened Disneyland in California when news of the Automatic Interplanetary Station broke, an unhappy irony for them.

Its purpose was not immediately obvious and news managers had decided that the objective of photographing the moon should not yet be mentioned, presumably in case

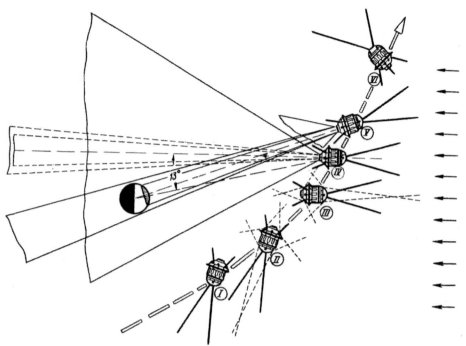

Trajectory of the Automatic Interplanetary Station, with key stages of orientation (I to VI)

of failure. They had good reason to be cautious, for confidence in the probe down on the ground was low, for signal transmissions from the probe were unreliable and those that were received indicated that it was overheating in the harsh conditions of Earth–moon space. The station reached a temperature of 40°C, far above that intended (25°C). The signals had become ever weaker and even with a dish ten times bigger than Kochka, Jodrell Bank had lost track, the British getting the impression that things had gone badly wrong.

Korolev at once flew with Mstislav Keldysh, Boris Chertok and other OKB-1 engineers from Moscow to see what could be done to salvage the situation. They rushed to Vnukovo Airport, the main domestic terminal in Moscow, where the government made available the fastest plane in the Aeroflot fleet, the Tu-104 jetliner. Such was the rush that the last passengers boarded as it taxied out to take off. Once they landed in the Crimea, a helicopter was supposed to bring them the rest of the way, but thick snow was falling, visibility was nearly zero and the helicopter had to fly on to Yalta. Here, local communist chiefs organized Pobeda cars to whisk them to Kochka where they eventually arrived, tired and probably worried sick. Korolev took charge, they worked through the night and by realigning the aerials ground control was able to send fresh commands up to the probe. Commands were sent up to change the spin rate and to shut some systems down and this had the desired effect of bringing temperatures down a bit, to 30°C. At about 65,000 km, rotation was stopped altogether.

The station swung around the south lunar pole at a distance of 6,200 km at 17:16 Moscow time on the 6th October before climbing high over the moon's far northern side. Now the sun angle was from behind and shining on the lunar farside. Early the following morning, the 7th October, rising 65,200 km above the moon's surface, sensors detected the sunlit farside of the moon and the Ye-2's unique design came into its own. The orientation system, linked to gas jets, went into action. One sensor locked onto the Sun, the other onto the moon. The gas jets fired from time to time to maintain this orientation. At 06:30 Moscow time, the camera system whirred into operation. For a full 40 min the two lenses took 29 pictures of the farside, with speeds varying between 1/200 and 1/800 sec. The last image was taken at an altitude of 66,700 km. The photographs were then developed, spooled, dried and scanned at 1,000 lines by the cathode ray television system. The system is not unlike a scanner that might be used on a modern domestic computer – except that this was 1959 and half a million kilometres away!

But how would ground control get the pictures? The station was transmitting during the picture taking, but the signal was intermittent and, to save energy, the transmitter was then turned off. Later that day, 7th October, the first attempt to send the images was made. One picture was received, taken some distance from the moon and showing it to be round, but not much more. Jodrell Bank picked up these signals, but – in order to take out radio noise so as to get a better signal – the station mistakenly cut out the video part of the signal.

The station was getting ever farther away on its elongated orbit. Near apogee, at 467,000 km, a second attempt was commanded to slow-transmit the pictures, but again the quality was very poor, so ground control just had to wait until its figure-of-eight trajectory brought the station back towards the Earth, which meant a long wait of almost two weeks. The Automatic Interplanetary Station's orbit took it far out behind the moon and it did not curve around back toward Earth until five days later, on 11th October, passing the moon's distance but this time Earthbound on the 15th. By the 17th, the station was halfway between Earth and the moon and it rounded Earth on the 19th. Now it was in an ideal position for the northerly Soviet ground-tracking stations.

Ground control made several attempts to get the probe to send the pictures, this time on fast speed. The first time, the next day, the signals were too weak. For the next four attempts, there was too much static and radio noise. In ground control, it became apparent that 29 pictures had indeed been taken, but whether they would ever receive them in useable condition was less clear. To lower the level of radio noise, the Soviet authorities ordered radio silence in the Black Sea and naval ships put out to sea off the Crimea to enforce the ban. The already tense humour in the control room became nervous and despondent. On the fifth attempt, though, the signal strength and quality improved abruptly. In the end, 17 of the 29 pictures were useable, covering 70% of the farside (the eastern side, as seen from Earth). On the 19th, rumours swept Moscow that pictures had been received of the farside of the moon.

Not until ten days later did the USSR release the historic first photograph of the moon's farside. A first set had been prepared by Yuri Lipsky in the Sternberg Astronomical Institute. The main picture was hazy and fuzzy, but it gave a bird's

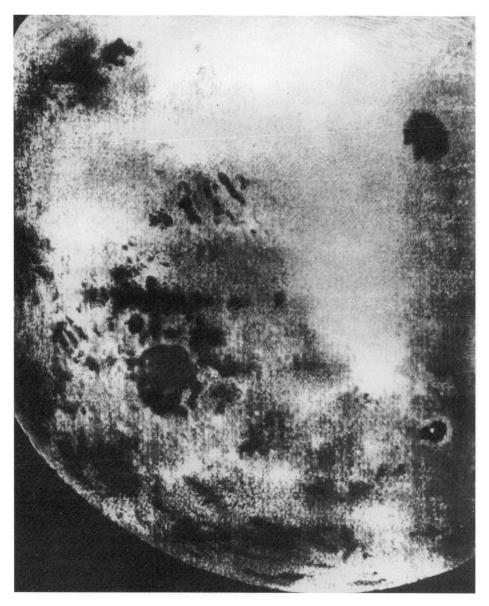

Automatic Interplanetary Station – first image

eye view of the moon's hidden side. It was the first time the view from space had ever been presented to people on Earth, the first time that a space probe had ever obtained data that could never have been obtained any other way. The farside was found to be mainly cratered highlands and was quite different from the near side. In the tradition of exploration, to the finder fell the privilege of naming the new-found lands. There

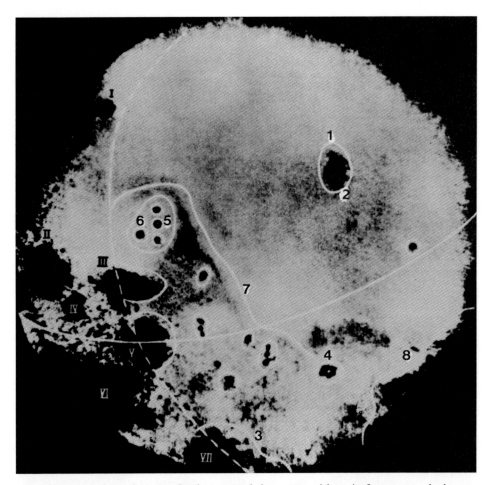

Automatic Interplanetary Station around the moon, with main features marked

was one huge crater, which the Russians duly called Tsiolkovsky and two seas, which they duly named the Moscow Sea (*Mare Moskvi*) and the Sea of Dreams. By astonishing coincidence, the pictures came through just when the monthly, popular live BBC astronomy television programme, *The sky at night* was on air (in those days, all programmes were done in real time with no pre-recording). The presenter, Patrick Moore, was able to show the pictures live to the world the instant they became available. Half a century later, he would still recall how the night the moon pictures came in was one of the highlights of his broadcasting career.

Contact with the Automatic Interplanetary Station was lost later that month, on the 22nd October. It passed the moon again on 24th January 1960, but signals could no longer be received. Its irregular orbit brought it crashing into the Earth's atmosphere at the end of April 1960, where it duly burned. For the Americans, the Automatic Interplanetary Station buried another myth: that the Russians could only build crude spacecraft on big dumb boosters. The station was a versatile display of engineering and technical sophistication. Now the whole world could see the pictures of the farside, be they in the newspapers or on educational posters. The Soviet Union published the first, primitive lunar farside atlas. Articles were published about the characteristics of the farside in general and of its specific features. A geological reconstruction was later made of the Moscow Sea [7].

Indeed, the Americans were so impressed with the Automatic Interplanetary Station that they contrived a plot of which James Bond and his director, M, would have been proud. In December 1959, only two months after the mission, the Russians sent a model of the station to an exhibition in Mexico. In reality, it was more than just a scale model, but the backup, working version. The Central Intelligence Agency sought and obtained the permission of the president of Mexico to kidnap the spacecraft. On its way to the exhibition, the truck carrying the spacecraft was diverted overnight to a timber warehouse where specialists were on hand to photograph, disassemble and reassemble the spacecraft. They had only a few hours to carry out their mission before anyone noticed that the truck was late. Although the main purpose was to estimate what size warhead the Soviet rocket could deliver, the exercise gave the Americans literally a hands-on examination of the capacity of Russian electronics, cameras and manufacturing capacity. The kidnapping of the Automatic Interplanetary Station was kept secret until the Cold War was long over.

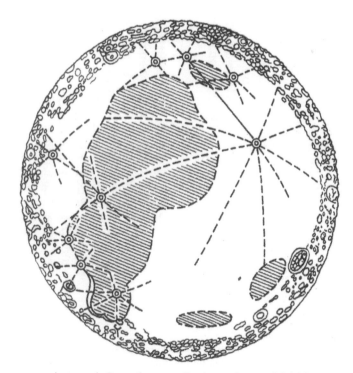

Automatic Interplanetary Station: scheme of farside

Instruments of the Automatic Interplanetary Station
Camera photography system, 200 mm and 500 mm.
Cosmic ray detector.
Micrometeoroid detector.

The original design for the Ye-2 series was based on two suitable launch windows: October 1959 and April 1960. The second window was now approaching. This time, the Russians would film the moon's farside while approaching the farside, covering the 30% not accessible to the AIS. Scientific Research Institute NII-380 devised an improved camera system and two probes were built, called the Ye-2f series. Now the earlier unreliabilities reasserted themselves. The first launching on 15th April began well, but Kosberg's RD-105 engine cut off early and the probe fell back from an altitude of 200,000 km. The second launching, the next day on 16th April, was even worse (some accounts give the date as 19th April). A moment after liftoff, the four strap-on blocks peeled apart, shooting out over the heads of the controllers, shattering the assembly hall and leaving the rails to the pad in a gnarled, tangled mess. None of this reached the rest of the world – or at least not for 30 years. So far as the rest of the world was concerned, the Russians had done one lunar farside mission and succeeded so completely that they did not need to repeat it.

FIRST MOONSHOTS

Soviet Union		United States	
17 Aug 1958	Cancelled (Ye-1)	17 Aug 1958	Failed after 77 sec
23 Sep 1958	Failed after 90 sec (Ye-1)	11 Oct 1958	Pioneer 1 (113,780 km)
12 Oct 1958	Failure after 104 sec (Ye-1)	7 Nov 1958	Pioneer 2 (1,549 km)
4 Dec 1958	Failure (Ye-1)	6 Dec 1958	Pioneer 3 (107,268 km)
2 Jan 1959	First Cosmic Ship (Ye-1)	3 Mar 1959	Pioneer 4 (passed moon)
18 Jun 1959	Failure (Ye-1a)	10 Sep 1959	Pad fire and explosion
9 Sep 1959	Pad abort (Ye-1a)	26 Nov 1959	Failed at 45 sec
12 Sep 1959	Second Cosmic Ship (Ye-1a)	26 Sep 1960	Fell back into atmosphere
4 Oct 1959	AIS (Ye-2a)		
15 Apr 1960	Failure (Ye-2f)	15 Dec 1960	Failed at 12 km
16 Apr 1960	Failure (Ye-2f)		

First moonshots: scientific outcomes
Failure to detect magnetic field, radiation belts around moon.
Lunar farside had few *mare*; mainly chaotic upland, one large crater.
Discovery and measurement of the solar wind.

This ended the early moon programme. Both countries experienced almost equally high launch failure rates. The difference was that two Russian probes fully succeeded in their missions. Because their failures were not known at the time and because the First Cosmic Ship was declared a success, they created an impression of unblemished competence. World opinion could see that the Soviet Union had taken an early and impressive lead on the path to the moon.

REFERENCES

[1] The person who has followed tracking issues concerning early Soviet lunar and inter-planetary probes is Grahn, Sven:
Mission profiles of early Soviet lunar probes;
Why the west did not believe in Luna 1;
Luna 3 – the first view of the moon's far side;
Soviet/Russian OKIK ground station sites;
The Soviet/Russian deep space network;
Jodrell Bank's role in early space tracking;
Yevpatoria – as the US saw it in the 60s at *http://www.users.wineasy.se/svengrahn/histind*
[2] Burchett, Wilfred and Purdy, Anthony: *Cosmonaut Yuri Gagarin – first man in space.* Panther, London, 1961.
[3] The person who has carried out the fundamental research into instrumentation on Soviet lunar and interplanetary probes is Mitchell, Don P. (2003–4):
– *Soviet space cameras;*
– *Soviet telemetry systems;*
– *Remote scientific sensors* at *http://www.mentallandscape.com*
[4] Caidin, Martin: *Race for the moon.* Kimber, London, 1959.
[5] Nesmyanov, A.: *Soviet moon rockets – a report on the flight and scientific results of the second and third space rockets.* Soviet booklet series #62, London, 1960; *Soviet planet into space.* Soviet booklet series #48, London, 1959.
[6] Lovell, Bernard:
The story of Jodrell Bank. Oxford University Press, London, 1968.
Out of the zenith – Jodrell Bank, 1957–70. Oxford University Press, London, 1973.
[7] Shevchenko, V.V.: *Mare Moskvi. Science and Life,* vol. 3, #88.

3

Planning the lunar landing

After the flight of the First and Second Cosmic Ship and then the Automatic Interplanetary Station, the full attentions of OKB-1 switched to manned spaceflight. Design of the first manned spaceship gathered pace over 1959, culminating in the launch of the first prototype, Korabl Sputnik, on 15th May 1960. The rest of the year was spent on refining the design, testing and preparing the first team of cosmonauts for flight. Only after five Korabl Sputnik missions, with dummies and dogs, were the Russians prepared to commit a cosmonaut to such a mission, called for the purpose the Vostok spacecraft. Vostok was a spherical spacecraft, riding on an equipment module, weighing four tonnes, able to fly a cosmonaut in space for up to ten days.

VOSTOK ZH: A CONCEPTUAL STUDY

This did not mean that moon plans were in abeyance, but they took a second place to the priority Korabl Sputnik and Vostok projects. Fresh Soviet moon plans were developed following the 5th July 1958 plan *Most promising works in the development of outer space*. Already, the designers were exploring how best to make a manned mission to the moon. In 1959, Sergei Korolev had asked Mikhail Tikhonravov and his Department #9 to work on the problems of rendezvous, using the now available R-7 launcher and to develop a broad range of missions before a heavy lift launcher, called the N-1, or Nositel ('carrier') 1 could be built. The department's initial design was to link a Vostok spacecraft, then being prepared for the first manned flight into space, with two or three fuelled rocket stages. The manoeuvrable, manned Vostok would carry out a number of dockings and assemble a complex in orbit. Once assembled, the rocket train would blast moonward. This has sometimes been called the Vostok Zh plan [1]. Such a flight would go around the moon, without orbiting or landing, flying straight back to Earth after swinging around the farside. Vostok Zh was a conceptual study and does not seem to have got much further. It was one of a number of possibilities explored during this period, the other principal one being a space station called Sever.

The limits of Vostok as a round-the-moon spaceship were realized at a fairly early stage. First, it was designed for only one person, while a moon mission required a crew of two, one as a pilot, the other as a navigator and observer. Second, a spherical-shaped cabin could only make a steep ballistic return into the Earth's atmosphere. The return speed from the moon was 11 km/sec, compared with 7 km/sec from Earth orbit and this would present difficult challenges to protect the cabin from the intense heat involved. Not only that, but an equatorial moon–Earth trajectory would bring the returning lunar cabin back to Earth near the equator. This was not a problem for the Americans, for they preferred sea splashdowns, but it was for the Russians, for no part of the Soviet Union was anywhere near the equator.

Looking for solutions to these problems, Tikhonravov collaborated with a rising engineer in OKB-1, Konstantin Feoktistov. He was a remarkable man. Born in Voronezh in 1926, he was a child prodigy and by the time the war broke out had mastered advanced maths, physics and Tsiolkovsky's formulae. When the Germans invaded, he acted as a scout for the partisans, but he was captured and put before a firing squad. The Germans left him for dead, but the bullets had only grazed his brain. He recovered, made his way back to the Russian lines, entered the Baumann Technical College in 1943, was awarded his degree and entered Mikhail Tikhonravov's design department in the 1950s.

Tikhonravov and Feoktistov worked to develop a spacecraft that could safely return to land following a high-speed reentry into the Earth's atmosphere from the moon. This led them away from the spherical shape of Vostok toward a headlight-

Konstantin Feoktistov

shaped acorn-like cabin. Tikhonravov calculated that coming through reentry at 11 km/sec the cabin could tilt its heatshield downward, use it to generate lift and skip across the atmosphere like a pebble skimming across water, bounce back into space and return to Earth, but now with a much diminished velocity [2]. This would not only reduce gravity forces for the crew, but make the capsule fly from the equator, skimming the atmosphere to a more northerly landing site in the Soviet Union. Although the return to Earth required considerable accuracy and although the reentry profile was a long 7,000-km corridor, it held out the promise of a safe landing on Soviet territory with a landing accuracy of ±50 km.

THE SOYUZ COMPLEX

These ideas were developed a stage further by another department of OKB-1, Department #3 of Y.P. Kolyako and by Korolev himself later in 1962. Korolev was already working on a successor to the Vostok spacecraft. Korolev's concept was for a larger spaceship than Vostok, with two cabins, able to manoeuvre in orbit and carry a crew of three. He linked his ideas to Tikhonravov's earlier concept of orbital assembly and approved on 10th March 1962 blueprints entitled *Complex for the assembly of space vehicles in artificial satellite orbit (the Soyuz)*, known in shorthand as the Soyuz complex. This described a multimanned spacecraft called the 7K which would link up in orbit with a stack of three propulsion modules called 9K and 11K which could send the manned spacecraft on a loop around the moon. This was endorsed by the government for further development on 16th April 1962. A second set of blueprints, called *The 7K9K11K Soyuz complex*, was approved on 24th December 1962. The complex was a linear descendant from the Vostok Zh design.

Work continued on refining the design of the Soyuz complex into the new year. On 10th May 1963, Korolev approved a definitive version, called *Assembly of vehicles in Earth satellite orbit*. The complex comprised a rocket block, which was launched 'dry' (not fuelled up) and which was the largest single unit. It contained automatic rendezvous and docking equipment and was labelled the Soyuz B; a space tanker, containing liquid fuel, called the Soyuz V; and a new manned spacecraft, called the Soyuz A. This was the system developed by Korolev, Tikhonravov and Feoktistov.

Soyuz A was a new-generation spaceship, 7.7 m long, 2.3 m diameter, with a mass of 5,800 kg. The design was radical, to say the least. At the bottom was an equipment section with fuel, radar and rocket motor. On top of this was a cone-shaped cabin for a three-man crew. Orthodox enough so far, but on top of that was a large, long cylinder-shaped orbital module. This provided extra cabin space (the cabin on its own would be small) and room for experiments and research. Work proceeded on the Soyuz complex into 1964 and a simulator to train cosmonauts in Earth orbit rendezvous was built in Noginsk.

Like the Vostok Zh, the Soyuz complex was aimed at flying cosmonauts around the moon without landing or orbiting. However, the Soyuz was large enough to carry two or even three men; had a cabin for observations and experiments; and the acorn

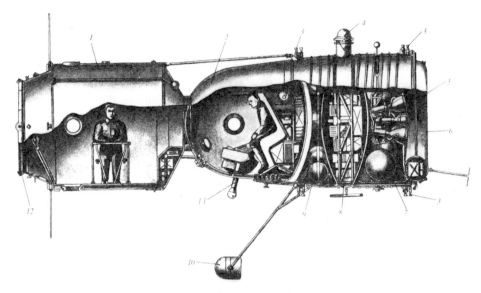

Soyuz complex – first block design (Soyuz A)

cabin that could tip its heatshield in such a way as to make a skip return to Earth possible.

The sequence of events for a moon flight was as follows. On day 1, the 5,700-kg rocket block, Soyuz B, would be launched into an orbit of 226 km, 65°. It would be tested out to see that its guidance and manoeuvring units were functioning. On day 2, the first of three 6,100 kg Soyuz V tankers would be launched. Because the fuel was volatile, it would have to be transferred quite quickly. The rocket block would be the 'active' spacecraft and would carry out the rendezvous and docking manoeuvres normally on the first orbit. Fuel would then be transferred in pipes. After three tanker linkups, a Soyuz A manned spaceship would be launched. It would be met by the rocket block, which, using its newly-transferred fuel, would blast moonwards. The on-going work was endorsed by government resolution on 3rd December 1963, which pressed for a first flight of the 7K in 1964 and the assembly of the Soyuz complex in orbit the following year. The first metal was cut at the very end of 1963 in the Progress machine building plant in Kyubyshev.

For the Soyuz complex, an improved version of the R-7 was defined. Glushko's OKB-456 was asked to uprate the RD-108 motor of block A and the RD-107 motors of blocks B, V, G and D, and gains of at least 5% in performance were achieved. In OKB-154, Semyon Kosberg also uprated the third stage. An escape tower was developed by Department #11 in OKB-1. After a number of evolutions, the new rocket was given the industry code 11A511. The improved motors were tested during 1962 and entered service over 1963–4.

The Soyuz complex lunar project was a complicated profile, involving up to six launches and five orbital rendezvous. Subsequent studies show that such a mission, assuming the mastery by the USSR of Earth orbit rendezvous, was entirely feasible [3].

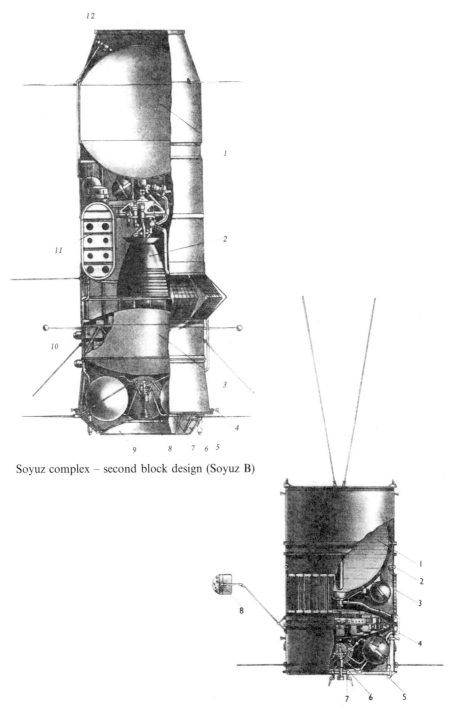

Soyuz complex – second block design (Soyuz B)

Soyuz complex – third block design (Soyuz A)

First flights were set for 1964 with the circumlunar mission for 1965–6. Had the Soviet Union persisted single-handedly with the design, then Russian cosmonauts could well have flown around the moon using this technique before the 50th anniversary of the revolution in October 1967.

Soyuz complex, 1962–4

Spacecraft	Soyuz A, B, V
Total mass	18 tonnes
Length	15m
Designer	Sergei Korolev

Soyuz A

Role	Manned spacecraft, <three cosmonauts
Weight	6.45 tonnes
Length	7.7 m
Diameter	2.5 m

Soyuz B

Role	Rocket block
Weight	5.7 tonnes
Length	7.8 m
Diameter	2.5 m

Soyuz V

Role	Tanker
Weight	6.1 tonnes
Length	4.2 m
Diameter	2.5 m

The Soyuz complex, using Earth orbit rendezvous (EOR) was a natural proposition for a nation bred on the theories of Tsiolkovsky. The two other possible methods of going to the moon were direct ascent and a much more obscure method called lunar orbit rendezvous (LOR). Direct ascent was the most popular one in the science fiction literature of the time. The *Stories of Tintin* cartoon is this type of method. A huge rocket – it really would have to be utterly enormous – would put up a moonship which would fly direct to the moon, slow down coming in to land, touch down and deposit two or three astronauts directly on the surface. After a period of exploration, the cosmonauts would climb back into their mother ship and fire direct back to Earth.

   Third, a variation on this was lunar orbit rendezvous (LOR). A booster would place both mother ship and lunar cabin directly into moon orbit, cutting out the Earth orbit rendezvous stage. The lunar cabin would descend to the surface while the mother ship continued to orbit. After surface exploration, the lander would take off, fly into lunar orbit and rendezvous with the orbiting mother ship. The lander crew would transfer to the mother ship before all the astronauts blasted out of lunar orbit for home. This method depended on a big and reliable booster, though nothing as big as direct ascent. Carrying out a rendezvous in distant lunar orbit was clearly a risky aspect of the plan. Alexander Shargei (AKA Yuri Kondratyuk) had outlined such a method, but again it depended on a rocket much bigger than anything immediately in prospect.

Origins of the manned Soviet moon programme, 1959–64

1959	Start of studies by Mikhail Tikhonravov in Department #9, OKB-1.
1962	Vostok Zh study.
1962	First design of the Soyuz complex (10th March).
	Endorsed by government (16th April).
	Second set of blueprints (24th December).
1963	Definitive design of the Soyuz complex (10th May).
	Approval by government (3rd December).

The design for the Soyuz complex required rendezvousing spacecraft to come within 20 km of one another on their first orbit so as to prepare for subsequent docking. This was something which the Vostok programme, limited though it was, could put to the test. On 11th August 1962, the third Vostok was put into orbit, manned by Andrian Nikolayev. Vostok 4 was put into orbit with Pavel Popovich exactly one day later, so precisely that it approached to within 5 km of Vostok 3 on its first orbit. This close approach was much better than anticipated. Both ships orbited the Earth together for three days, though unable to manoeuvre and drifting ever farther apart. In June 1963, Vostok 6, with Valentina Terreskhova on board, came to within 3 km of Vostok 5. Even though there was never any prospect of the ships coming together, the two group flights were a demonstration of how close spaceships could come on their first orbit. The cosmonauts communicated with one another during their missions and ground control learned how to follow two missions simultaneously.

Although these missions had been put together at relatively short notice and in an unplanned way to respond to the flights of the American Mercury programme, this was not at all how the missions had been interpreted in the West. The Vostok missions were seen as a carefully orchestrated series of events leading up to a flight to the moon. When Pavel Popovich joined Andrian Nikolayev in orbit, the Associated Press speculated that an attempt might be made to bring the spacecraft together before setting out on a loop to the moon. It was almost as if the agency had seen the designs of the Soyuz complex.

During the 1963 conference of the International Astronautical Federation in Paris, Yuri Gagarin told the assembled delegates:

A flight to the moon requires a space vehicle of tens of tonnes and it is no secret that such large rockets are not yet available. One technique is the assembly of parts of spaceships in near-Earth orbit. Once in orbit the components could be collected together, joined up and supplied with propellant. The flight could then begin.

This was not how the Americans were planning to go to the moon – NASA had opted for Shargei's LOR method – and many people were skeptical as to how truthful Yuri Gagarin was actually being. *The Russians must be racing the Americans to a moon*

Nikolayev, Popovich return in triumph

landing, they said. In fact, Gagarin was outlining, perfectly accurately, the Soviet moon plan as it stood in autumn 1963.

JOINT FLIGHT TO THE MOON PROPOSED BY KENNEDY, AGREED BY KHRUSHCHEV

The Soyuz complex remained the Soviet moon plan until August 1964 and, as can be seen, was limited to a manned circumlunar mission. A man-on-the-moon programme was not even on the drawing board. A central assumption in NASA, in the American political community and in the Western media was that the Soviet Union had a long-time plan to send a man to the moon. All Soviet missions were explained in the context of this presumed, methodical master plan. To the West, it was unthinkable that the Soviet Union was not trying to plant the red flag on the moon first.

In reality, until August 1964, the Soviet Union had no such plan at all. The significance of the Kennedy challenge of 25th May 1961 had not been fully appreciated by the Kremlin. Kennedy's speech may have been considered aspirational or rhetorical, rather than the purposeful mobilization of an entire national effort that it became. For the Soviet Union, well used to public hyperbole, it was just another speech. According to his son, Sergei, Nikita Khrushchev 'did not attach much importance to the challenge of John F. Kennedy'. But as time went on and the American space industry burgeoned, he was now faced with the choice of accepting the challenge and spending billions, or allowing his richer competitor to get ahead. 'My father was not prepared to answer the question and neither was Korolev', his son said later. Khrushchev had a sense of financial proportion lacking in his successor, wanted to concentrate the Soviet Union's limited resources on housing and agriculture and was savvy enough to know that for the Americans the Apollo programme was almost small change. Sergei Khrushchev said: 'It had never been my father's plan to spend substantial sums in order to support our priority in space'.

It is also worth recording that, although the American moon project appeared, in retrospect, to be purposeful and deliberative, this was not always how it seemed at the time. President Kennedy had accepted the project somewhat reluctantly, had cast around for alternatives, but had been hustled into the venture by his vice-president and space enthusiast Lyndon Johnson. Following his speech to Congress in May 1961, only five congressmen had spoken and the project had been nodded through. There was little sign of public enthusiasm. Congress had approved many projects before and they had not materialized: the moon programme could well have gone their way too. Kennedy himself seems to have been uncertain and when he met Nikita Khrushchev in Vienna later that June, Kennedy proposed a joint lunar venture. According to Sergei Khrushchev, this proposal 'found my father unprepared'. Nikita Khrushchev did not formally respond. His own military were unhappy about sharing their rocket secrets (so was Korolev) and Khrushchev was concerned that the Americans would find out how weak their missile forces really were. Khrushchev baulked at Korolev's estimates for the cost. The Russians were probably wary of exposing just how unprepared they were for such a venture.

The situation changed two years later. Kennedy repeated his proposal, this time to Soviet ambassador Anatoli Dobrynin in August and then to the United Nations in September. Many query his motives. Some say that Kennedy made the offer with the certainty that the Soviet Union would refuse, but make him look statesmanlike to the world. Privately, some people on the American side were apprehensive about sharing with the Russians, as they were in turn.

Whatever these convoluted manoeuvrings, Khrushchev was more positive this time. The Soviet Union now had its own fleet of R-16 missiles and was closer to strategic military parity. With the Cuban misadventure left behind, he felt he could now work with Kennedy. The prospect of avoiding an expensive moon race and actually sharing costs began to look attractive. In a compromise with the military, a joint moon programme would be done on the basis of each partner doing its share independently, minimizing the risk of Russia learning of American technical capacities. In the autumn of 1963, Nikita Khrushchev volunteered that the Soviet Union was not in a moon race and had no plans to send men to the moon. On 26th October, he declared:

The Soviet Union is not at present planning flights of cosmonauts to the moon. Soviet scientists are studying it as a scientific problem.

The Americans want to land a man on the moon by 1970. We wish them luck and we will watch to see how they fly there and how they will return. I wish them success. Competition would not bring any good but might to the contrary cause harm because it might lead to the death of people.

He opened the door to the enterprise being undertaken together. Now that Russia had launched a woman into space, he jokingly referred to how an American man could fly to the moon with a Soviet woman, but then countered by saying the gender balance should be reversed and a Russian man should bring an American woman there. This was the second round of what could have turned into a courtship for a joint pro-gramme. There was a positive response from the White House which said it was studying the premier's statement. Although Khrushchev had now responded to the Vienna offer two years late, he had now responded positively. We know of no technical documents that could have outlined how such a joint venture might have taken place. The proposal came to an abrupt end only three weeks later when John Kennedy was assassinated in Dallas, Texas, where he arrived directly after visiting the growing space facilities of Houston, Texas. The new leader, Lyndon Johnson, had built his political reputation on responding to the Soviet threat in space and made it clear that he was not interested in a joint programme.

Thus by early 1964, the Soviet Union:

- Had its own plan to fly to the moon, the Soyuz complex, now in construction.
- Based its approach on Earth orbit rendezvous.
- Planned only a flight around the moon, without orbiting or landing.
- Had paid little or no attention to Kennedy's speech of 25th May 1961.
- Was reluctant, principally for reasons of cost, to compete with the United States.
- Had begun, slowly, to respond to overtures for a joint lunar venture with the Americans.
- But now found the Kennedy idea rejected by the new American president.

REVISING THE SOYUZ COMPLEX

In early 1964, Russia's plans were still to fly around the moon using the Soyuz complex. With the construction of Soyuz already under way, the R-7 rocket already available and the first group flights showing remarkable promise, there was a real prospect that this could be achieved over 1966–7 or so.

There was still considerable uncertainty about the future medium- and long-term direction of the Soviet space effort. The death of John Kennedy had now eliminated the prospect of a joint mission. In 1963, Jodrell Bank Observatory director Bernard Lovell had visited the Soviet Union as a guest of Mstislav Keldysh and learned, to his surprise, that the Soviet Union had no plans to race the Americans to the moon (exactly as Khrushchev had told the United Nations). Instead, they would build an Earth-orbiting space platform. Indeed, designs of Soviet cosmonauts spacewalking around such platforms soon found their way to the West. Bernard Lovell's remarks were disputed by some Soviet scientists, but his visit created some considerable doubt about the nature of Soviet intentions.

Although the Soyuz complex had made considerable progress during 1962–3, this slowed down during 1964. However, it is important to stress that the Soyuz complex was no mere study. Not only did the design progress to an advanced stage, but initial flight models were in construction. The slowdown was not because of an action on the part of government, but due to gross overwork in OKB-1. Concerned with the complexities of the Earth orbit rendezvous manoeuvres required, Korolev now began to revise the concept. The weight of the complex to be assembled in Earth orbit would be about the same, 18 tonnes. Under the new plan:

- Only three spacecraft would be involved.
- The rocket block would use the much more powerful hydrogen fuel.
- The Soyuz spacecraft would, for the lunar journey, be shortened and lightened to five tonnes: the orbital module would not be carried. This would now be called the Soyuz 7K-L-1 (L for Luna, *Luniy* or moon).

Learning about this, a rival design bureau, OKB-52 of Vladimir Chelomei, came up with a rival proposal. Using the new Proton rocket which he was building, he said that he could send such a spacecraft directly to the moon. Only one rocket would be required and there was no need for orbital rendezvous or the transfer of fuels in Earth orbit. He persuaded the government that the plans for Earth orbital rendezvous were too cumbersome. Korolev was so busy with other projects and Chelomei managed to get government approval before he realized what was going on and could stop him.

The arrival of a competitor to Korolev was an important development. Until 1964, Korolev had, as chief designer, ruled supreme over the Soviet space programme. Vladimir Chelomei was a slightly younger man than Korolev – he was born in 1914 – and when Korolev had developed the German V-2 after the war, Chelomei had built derivatives of the V-1 flying bomb. From 1944 to 1954, Chelomei had developed pulse jet engines, cruise missiles and sea-borne rockets. His style was quite different from Korolev, being smartly dressed, with a polished manner and he was a great communicator. All who met him paid tribute to his ambition and powers of persuasion. Chelomei was a professor of the Baumann Technical School, a member of the Academy of Sciences from 1958 and full academician from 1962. He was able to offer the Kremlin a viable military space programme: new military rockets (SS-9, Tsyklon, Proton), anti-satellite weapons (Polyot), radar observation satellites and was even working on a manned platform for space surveillance (Almaz). Nikita Khrushchev's son Sergei worked for him.

Chelomei was not the only challenger to Korolev's hitherto undisputed prominence. Korolev's former collaborator, Valentin Glushko, ran a large engine design bureau, OKB-456, and as we saw in 1958 the two had already quarrelled over the upper stage for the R-7 used to fire the first cosmic ship. In Dnepropetrovsk, Ukraine, another large design bureau had grown up under Mikhail Yangel. He built military missiles for the Soviet rocket troops by the hundreds in his sprawling factory there.

Vladimir Chelomei

Some of the missiles were adapted as satellite launchers and by 1962 his design bureau was building small military satellites.

SOVIET DECISION TO GO TO THE MOON, AUGUST 1964

The Soviet decision to land on the moon was not made until August 1964, more than three years after Kennedy's address to Congress. Examination of the Soviet documentary record in the 1990s suggests that as 1963 turned to 1964 there was a dawning realization of the scale of the American commitment under Apollo. Soviet intelligence reported on the burgeoning American effort, though there was no need to rely on spies, for the American programme was enthusiastically publicized in the open literature. Soviet designers put it up to their own leadership that they had to respond. Again, the decision was taken as a result of pressure from below, rather than because of a government *diktat* from on high. Until spring 1964, the Soviet space programme had largely been shaped by goals set by Korolev, Tikhonravov and others in proposals and memoranda outlining a step-by-step *Russian* approach to space exploration. Now, a subtle shift occurred, with Soviet goals now determined in respect of *American* intentions.

The process of reappraisal began in the course of 1963. That autumn, Korolev restated and revised his approach, presenting a fresh set of plans to government in which he outlined how Soviet lunar exploration should progress. This was *Proposal for the research and familiarization of the moon*, by Sergei Korolev on 23rd September 1963. They were all labelled L- after the Russian word for moon:

L-1   Circumlunar mission using the Soyuz complex.
L-2   Lunar rover to explore landing sites.
L-3   Manned landing.
L-4   Research and map the moon from orbit.
L-5   Manned lunar rover.

What is interesting here is the prominence given to a manned *landing*, which had hitherto not featured at all in Soviet planning. Khrushchev received representations from Chelomei, Yangel and Korolev that each one of them had the project that could respond to Apollo:

- Korolev offered the latest version of the Soyuz complex for a round-the-moon mission. He also had a powerful, heavy-lift N-1 booster under development, which could put a man on the moon. The project had developed only slowly since 1956 and was now languishing.
- Chelomei proposed his UR-500 Proton rocket for a direct around-the-moon mission and a much larger derivative, the UR-700 for a direct ascent lunar landing.
- Mikhail Yangel's bureau offered a third rocket, the R-56.

Siddiqi has chronicled how the Soviet approach changed in the course of 1964 [4]. The first American hardware had begun to appear and the Saturn I had begun to make its first flights. The various design bureaux saw the moon programme as a means of keeping themselves in business – and making sure that rivals did not rise to prominence at their expense. Korolev even made a blatant appeal to Khrushchev to the effect that it would be unpatriotic and unsocialist to let the Americans pass out Soviet achievements. Khrushchev eventually gave in and by this time the leading members of government, the party, the military and the scientific establishment had come round to the view that it would be wrong not to beat the Americans to the moon. A final contributory factor was that the Soviet Union had coasted through the successes of Gagarin, Titov and the two successful group flights. At some stage, the political leadership realized that complacency was no match for some serious forward planning.

Whatever the mixed circumstances, the government and party issued a resolution on 3rd August 1964, called *On work involving the study of the moon and outer space*. This resolution:

- Formally committed the Soviet Union to a moon-landing programme.
- Charged the task to Korolev's OKB-1, with the objective of landing a man on the moon in 1968. The N-1 heavy lift rocket, now eight years in design would be used.
- Committed the Soviet Union to continue to pursue the around-the-moon project. This would be done by Chelomei's OKB-52, with the objective of sending a man around the moon in 1967. This plan replaced the Soyuz complex.

This is one of the most important government decisions in our story. It was a joint party and government resolution, #655-268 to be precise. It gave the two bureaux the authority to requisition resources to bring these programmes to fulfilment. A word of caution though: although the party and government issued the decree, it was a secret one. Whilst known to the senior ranks of party, government and industry, it was not on the evening television news and indeed it was not uncovered until the Soviet Union had ceased to be.

The resolution was problematical for a number of other reasons. First, it came more than three years after the American decision to go to the moon, so the Russians were starting from far behind and also committed themselves to the finishing line sooner. Second, they divided the project into two distinct tasks, unlike the Americans who aimed to circle the moon on the way to a landing. The two tasks were given to two different design bureaux, meaning two different sets of hardware. The decision was a political compromise, giving one project to Korolev (at the expense of Yangel) and one to Chelomei (at the expense of Korolev). This might have been acceptable if the USSR had considerably more resources than the United States, but the very opposite was the case. Third, as we shall see, the Russians had a lot of difficulty in even keeping to the plans that were formally agreed. Fourth, it meant that Soviet methods of space exploration were determined less by the setting of objective goals and methods, but by reference to American intentions and the need to reach acceptable compromises between the ambitious design bureaux within the Soviet Union itself. Indeed, under Leonid Brezhnev, the Soviet system became less and less able to take hard choices, less able to say 'no', permitting and funding the many rival projects of the competing military–industrial élites simultaneously [5]. So the 1964 resolution was a pivotal, but problematic decision.

The original Soyuz complex was now gone from the moon plans, with the danger that four years' design work would now go to waste. Korolev saved the 7K spacecraft and made the case to the government that it should be adapted for Earth orbital missions and to test out rendezvous and other techniques that would be required for the moon landing. The 7K was now renamed the 7K-OK (OK standing for orbital craft, *Orbitalny Korabl*). The spacecraft was now called Soyuz, even though it had been one part of a much bigger project called the Soyuz complex. As such, it became the basis for the spacecraft still operating today. The intention was that the 7K-OK follow as soon as possible from the Vostok programme. In the event, Soyuz was delayed, had a difficult design history and did not make its first unmanned flight until 1966.

Thus in August 1964, the Soviet Union:

- Abandoned Earth orbit rendezvous as a means of flying a cosmonaut to the moon, scrapping the Soyuz complex.
- Matched President Kennedy's challenge to land an American on the moon by a commitment to land a Soviet cosmonaut there in 1968.
- Set the objective of sending a cosmonaut around the moon first, using the new Proton rocket and the skills of the Chelomei design bureau, in 1967.

With an economy half the size of the United States, the Soviet Union had set itself some daunting goals. Not only was it beginning the race three years after the United States, but it set itself an extra circuit to run – and still win both races a year earlier than its rival.

As part of the shake-out of 3rd August 1964, Tikhonravov's Department #9 in OKB-1 was disbanded. All the work it had done on orbital stations was transferred to the Chelomei OKB-52 for his programme for space stations, called Almaz. Little more was heard of Mikhail Tikhonravov, the father of the Soviet lunar programme, from there on. He was 64 years old then and appears to have retired at this point. Mikhail Tikhonravov eventually passed away aged 74 on 4th March 1974. His prominent role had been obscured by Korolev. It probably should not have been, for the Soviet state did honour this shy man with the Lenin Prize, two Orders of Lenin, 'honoured scientist of the Russian Federation' and the title 'Hero of socialist labour'. In a space programme dominated by giant egos, Mikhail Tikhonravov had been content to labour in the background, though he was never afraid to put forward proposals if that would advance the concepts and ideas he believed in so greatly. He never attracted or sought attention the way others did, but his influence on the Soviet lunar programme can only be considered profound, shaping all its early stages.

CHANGING WAYS TO GO

The 3rd August 1964 resolution *On work involving the study of the moon and outer space* should have settled the Soviet moon plan. On the surface of things, it not only set the key decisions (lunar landing, around the moon) but the method all in one go. By contrast, the Americans had decided how to go to the moon in two stages, taking the decision in May 1961 and settling on the method, LOR, in autumn 1962.

In reality, the decision of August 1964 settled much less than it appeared. Many of the parties involved continued to fight for the decisions of August 1964 to be remade. Korolev would not accept the allocation of the around-the-moon project to Chelomei and spent much of 1965 trying to win it back to his own design bureau, with some success. For his part, Chelomei began to present the UR-700 as an alternative to the rocket designated for the moon landing, Korolev's N-1.

Mikhail Tikhonravov at retirement

Whereas the Americans had debated between Earth orbit rendezvous, lunar orbit rendezvous and direct ascent, the debate in Russia was over which rocket to use: Korolev's N-1; Chelomei's UR-700; or Mikhail Yangel's R-56. Despite the government decision of August 1964, these were still in contention.

Russia's three ways to go

Korolev design bureau (OKB-1)	N-1
Chelomei design bureau (OKB-52)	UR-700
Yangel design bureau (OKB-586)	R-56

Korolev's N-1

Korolev had originally planned the N-1 as a rocket which would send large spaceships unmanned, then manned, on a flyby of Mars. The concept of the N-1 dated to the period 1956–7 and was refined over the next number of years by Mikhail Tikhonravov, Gleb Yuri Maksimov and Konstantin Feoktistov. Whereas the R-7 could lift four tonnes into Earth orbit and was a huge advance in its day, the N-1 was designed as a great leap forward to put 50 tonnes into orbit. Early designs assumed that the N-1 would be used for the assembly of a manned Mars expedition in Earth orbit. This would be for a Mars flyby, rather than a landing, much like Korolev's early designs for the moon. The 50 tonnes were gradually revised upward to 75 tonnes. Several such Mars proposals were developed in OKB-1 over 1959–67, based first around the assembly of 75-tonne interplanetary spaceships in Earth orbit [6].

N-1 was now adapted for a manned flight to the moon, though designers kept, in their bottom drawer, plans to redevelop the N-1 for a Mars mission, the N-1M. Korolev completed his design for the lunar N-1 on 25th December 1964.

The N-1 concept was reshaped around lunar orbit rendezvous, the same technique as that used by the Americans, although there were some differences in the precise detail. In the early stages, a double N-1 launch was considered necessary, with Earth orbit rendezvous preceding the flight to the moon, but this was seen as too complex, not essential and was eventually dropped. The tall N-1 was similar in dimensions to the American Saturn V, being almost exactly the same height. Unlike the Saturn V, the N-1 used conventional fuels (liquid oxygen and kerosene), which required a large number of engines of modest thrust, 30 altogether. The performance of the N-1 was inferior, able to send only two men to the moon and put only one on its surface.

R-56

The full designs of the UR-700 and the R-56 have not been fully revealed, though it is now possible to speculate with accuracy what they may have been like [7]. Like the N-1, the R-56 offered a minimalist lunar mission, with a lunar-bound payload of 30 tonnes. Chief designer was Mikhail Yangel. Born in Irkutsk on 25th October 1911 (os), he was a graduate of the Moscow Aviation Institute and after the war worked in Korolev's OKB-1. In 1954, he was given his own design institute, OKB-586 in Dnepropetrovsk. A model of the R-56 now appears in the company museum and sketches have been issued. The R-56 was a three-stage rocket, 68 m long, each cluster 6.5 m in diameter. To get from the Ukraine to Kazakhstan, it would be transported by sea – but this could be done on barges from the Dnepropetrovsk factory on the inland waterway system of the Soviet Union via the Syr Darya to Tyuratam.

The principal difference between the N-1 and the R-56 was the use of engines. For the R-56, Valentin Glushko's OKB-456, the old Gas Dynamics Laboratory, developed large, high-performance engines called RD-270. For many years, it had been assumed that the Soviet Union had been unable to develop such engines, but this was not the case. Unlike the American engines, which used liquid oxygen and liquid hydrogen, Glushko used storable fuels. His engines used unsymmetrical dimethyl methyl hydrazine (UDMH) and nitrogen tetroxide, producing a vacuum thrust of between 640 and 685 tonnes, a specific impulse of 322 sec and a pressure of 266 atmospheres in its combustion chamber. Each RD-270 weighed 4.7 tonnes, was 4.8 m tall and could be gimballed. Valentin Glushko managed to build 22 experimental models of the RD-270 and 27 firings were carried out in the course of October 1967 to July 1969, all showing great promise. Three engines fired twice and one three times.

Mikhail Yangel

RD-270 engine
Length	4.85 m
Diameter	3.3 m
Pressure	266 atmospheres
ΔV	3,056
Specific impulse	322 sec
Weight	4.77 tonnes (dry)
	5.6 tonnes (fuelled)

In the event, the R-56 did not turn out to be a serious competitor. There is some suggestion that Yangel saw the damage being done by the rivalry within the Soviet space industry and did not wish to press a third project that would divert resources even further. Authority to develop the R-56 seems to have been given by the government in April 1962, but a subsequent government decision in June 1964 ordered a cessation of work.

UR-700

More is known of the UR-700 [8] and in recent years the managers of the Konstantin Tsiolkovsky Museum in Kaluga helpfully put a model on display. The UR-700 would have a thrust of 5,760 tonnes, able to put in orbit 151 tonnes, a much better performance than the Saturn V. It would have been a huge rocket at take-off: 74 m tall, 17.6 m in diameter and a liftoff weight of 4,823 tonnes [9].

The UR-700 combined a mixture of strap-on rockets and fuel tanks (like the Proton) clustered around the core stage, with the three strap-ons jettisoned at 155 sec and the three core engines burning out at 300 sec. It was a typically ingenious Chelomei design, one building on the proven engineering achievement of the Proton. As for power, the engine in development for the R-56, the RD-270, was transferred from the R-56 to the UR-700. Chelomei's UR-700 had a single third stage, an RD-254 engine based on the Proton RD-253.

The UR-700 was a direct ascent rocket, which Chelomei believed was safer than a profile involving rendezvous in lunar orbit. Outlines of the UR-700 moonship are available. These were for a 50-tonne cylindrical moonship with conical top entering lunar orbit, 21 m long, 2.8 m in diameter, with a crew of two. The moonship would descend to the lunar surface backwards, touching down on a series of six flat skids. The top part, 9.3 tonnes, would blast off directly to Earth, the only recovered payload being an Apollo-style cabin that Chelomei later developed for his Almaz space stations. In his design, Chelomei emphasized the importance of using multiply redundant systems, the use of N_2O_4/UDMH fuels, exhaustive ground testing and the construction of all equipment in the bureau before shipping to the launch site. One reason for its slow pace of development was Chelomei's concentration on intensive ground testing [10].

Like the R-56, the UR-700 was proposed as a moon project before the decision of August 1964. The N-1 was made the approved man-on-the-moon project in August 1964 and so, in October 1964, the UR-700 was cancelled and, as we saw, work on the R-56 was also terminated. Never one to give up, Vladimir Chelomei continued to advocate his UR-700 design, even getting approval for preliminary design from the Space Ministry in October 1965, much to Korolev's fury. The following year, Chelomei got as far as presenting designs showing how the N-1 pads at Baikonour could be converted to handle his UR-700. Chelomei formally presented the UR-700 to a government commission in November 1966 as an alternative, better moon plan than the N-1. The government politely agreed to further research on the UR-700 'at the preliminary level' (basic research only) and this was reconfirmed in February 1967. Unfazed by this, Chelomei's blueprints for the UR-700 were signed on 21st July 1967, approved by party and government resolution # 1070-363 on 17th November 1967, three years *after* the N-1 had been agreed as the *final* moon design!

Designs for the UR-700 moonship were finalized on 30th September 1968. First launch was set for May 1972 and after a successful second unmanned flight, the third would have a crew (similar to the American Saturn V) with lunar landings in the mid-1970s. Although a certain amount of work was done on the project in 1968, it is unclear if much was done thereafter and it does not seem as if any metal was cut. The programme was not finally cancelled until 31st December 1970. In fairness to Chelomei, he never claimed, at least at this stage, that his UR-700 could beat the Americans to the moon. It is possible he saw the UR-700 as a successor project to the

N-1, or one that could later be adopted if the N-1 faltered. The UR-700 plan certainly had many fans, quite apart from Chelomei himself, believing it to be a much superior design to the old and cumbersome N-1. However, its reintroduction into the moon programme in late 1967 was yet another example of the rivalry, disorder, waste and chaos enveloping the Soviet moon programme.

UR-700

1st stage

Length	33.5 m
Diameter at base	15.6 m
Weight (dry)	146 tonnes
Engines	Eight RD-270

2nd stage

Length	18.5 m
Weight (dry)	49 tonnes
Engines	Four RD-270

3rd stage

Length	13.5 m
Engine	RD-210
Burn time	217 sec
Weight (dry)	16 tonnes

4th stage

Length	8.5 m
Diameter	4.15 m
Engines	Four RD-210
Weight (dry)	5.6 tonnes

Contrary to Western impressions that the Soviet space programme was centralized, in fact it operated in a decentralized, competitive way. Thus, in the period after the government decision of 1964, three design bureaux were at work not only designing but building rival moon projects. Again, this marked a key difference from the American programme. In the United States, rival corporations submitted proposals and bids, but only one was chosen to develop the project and build the hardware (the company concerned was called the prime contractor).

In the Soviet Union, by contrast, rival design institutes not only designed but built hardware. Decisions about which would fly were taken much later. As a result, the Soviet moon programme, and indeed other key programmes, contained several rival, parallel projects. This was something neither appreciated nor imagined to be possible in the West at the time. The rivalry between designers was at a level that could not have been conceived on the outside. At one stage, no less a person than Nikita Khrushchev tried to mediate between Sergei Korolev and Valentin Glushko, inviting the two to his summer house for a peace summit (he was not successful).

THE FINAL ROUTE DECIDED

Thus the resolution of August 1964 was much less decisive than one might expect. Not only did it not resolve the rivalry between different projects, but it did not ensure the rapid progress of those that were decided. Progress on the moon plan between August 1964 and late 1966 was quite slow. Not until October 1966 were steps taken to accelerate the favoured programme, the N-1, with the formation of the State Commission for the N-1, also known as the Lunar Exploration Council.

In September 1966, a 34-strong expert commission was called in to review the moon programme, decide between the N-1 and UR-700 and settle the continued rivalries once and for all. Mstislav Keldysh was appointed chairman and it reported at the end of November. Despite impressive lobbying efforts by Chelomei and Glushko to replace the N-1 with the UR-700, the original plan won the day. The commission's report, confirming the N-1 for the moon landing and the UR-500K for the circumlunar mission was ratified by the government in a joint resolution on 4th February 1967 (*About the course of work in the creation of the UR-500K-L-1*), which specified test flights later that year and a landing on the moon in 1968. The joint resolution reinforced the August 1964 resolution and upgraded the landing on the moon to 'an objective of national significance'. This meant it was a priority of priorities, enabling design bureaux to command resources at will. The real problem was that the Americans had decided on their method of going to the moon five years earlier and Apollo had been an objective of national importance for six years. In effect, the February 1967 resolution hardened up on the decision of August 1964. The Russian moon plan was now officially set in stone (though, in practice, the UR-700 was not finally killed off for another three years). The Keldysh Commission of 1966 and the resolution of 1967 would have been unnecessary had not the rival designers continually tried to re-make the original decision. It was a dramatic contrast to the single-mindedness of the Apollo programme and the discipline of American industry.

There were considerable differences between how the Russians and Americans organized their respective moon programmes. In the United States, there had indeed between intense rivalries as to which company or corporation would get the contract for building the hardware of the American moon programme. Once decisions were made, though, they were not contested or re-made and rival programmes did not proceed in parallel. In the United States, the decision as to how to go to the moon was the focus of intense discussions over 1961–2. No equivalent discussion took place in the Soviet Union. Until 1964, Earth orbit rendezvous, using the Soyuz complex to achieve a circumlunar mission, was the only method under consideration. When the N-1 was adopted as the landing programme in August 1964, lunar orbit rendezvous was abruptly accepted as the method best suited to its

Vladimir Chelomei and Mstislav Keldysh

dimensions, despite the investment of three years of design work in the Soyuz complex then reaching fruition.

Chelomei's UR-700 was a direct challenge to this approach and Chelomei raised questions about the risks involved in lunar orbit rendezvous. However, the debate in the Soviet Union was less about *how* to go to the moon, but, instead: *which bureau, which rocket, which engines* and *which fuels?*

RUSSIA'S THREE WAYS TO GO

	N-1	R-56	UR-700
Designer	Korolev	Yangel	Chelomei
Bureau	OKB-1	OKB-586	OKB-52
Method	LOR	LOR	Direct ascent
Height	104 m	68 m	74 m
Weight	2,850 tonnes	1,421 tonnes	3,400 tonnes
Moonship	33 tonnes	30 tonnes	50 tonnes
First-stage engines	NK-31	RD-270	RD-270

Key government and party decisions in the moon race

3 Aug 1964	*On work involving the study of the moon and outer space.*
16 Nov 1966	Keldysh Commission.
4 Feb 1967	*About the course of work in the creation of the UR-500K-L-1.*

Thus, by now, the Soviet Union had made a plan for sending cosmonauts around the moon and a separate plan for landing on it. A plan had been worked out for both missions. Hard work lay ahead in constructing the rockets, the spacecraft, the hardware, the software, the support systems and in training a squad of cosmonauts to fly the missions. In the meantime, unmanned spacecraft were expected to pave the way to the moon.

REFERENCES

[1] Wachtel, Claude: Design studies of the Vostok Zh and Soyuz spacecraft. *Journal of the British Interplanetary Society*, vol. 35, 1982.

[2] Hall, Rex D. and Shayler, David J.: *Soyuz – a universal spacecraft*. Springer/Praxis, Chichester, UK, 2003.

[3] Clark, Phillip S. and Gibbons, Ralph: The evolution of the Soyuz programme. *Journal of the British Interplanetary Society*, vol. 46, #10, October 1993.

[4] Siddiqi, Asif: *The challenge to Apollo*. NASA, Washington DC, 2000.

[5] Sagdeev, Roald Z.: *The making of a Soviet scientist*. John Wiley & Sons, New York, 1994.

[6] Zak, Anatoli: Manned Martian expedition. *http://www.russianspaceweb.com*, 2001.

[7] Clark, Phillip S.: The history and projects of the Yuzhnoye design bureau. *Journal of the British Interplanetary Society*, vol. 49, #7, July 1996.

[8] Sokolov, Oleg: The race to the moon – a look back from Baikonour. American Astronautical Society, *History* series, vol. 23, 1994.

[9] Lardier, Christian: Les moteurs secrets de NPO Energomach. *Air and Cosmos*, #1941, 18 juin 2004; H. Pauw: Support pulled on lunar mission. *Spaceflight*, vol. 45, #7, July 2003.

[10] Vick, Charles P.: The Mishin mission, December 1962–December 1993. *Journal of the British Interplanetary Society*, vol. 47, #9, September 1994.

4

The soft-landers and orbiters

With man-on-the-moon plans in full swing, the next stage for the Soviet Union was to send unmanned probes to pave the way. These were essential for a manned landing on the moon. The successful landing of a probe intact on the lunar surface was necessary to test whether a piloted vehicle could later land on the moon at all. The nature of the surface would have a strong bearing on the design, strength and structure of the lunar landing legs. The level of dust would determine the landing method and such issues as the approach and the windows. The successful placing of probes in lunar orbit was necessary to assess potential landing sites that would be safe for touchdown and of scientific interest. Stable communications would also be essential for complex operations taking place 350,000 km away. Unmanned missions would address each of these key issues, one by one.

ORIGINS

As noted in Chapter 2, the Soviet pre-landing programme can be dated to the 5th July 1958 when Mikhail Tikhonravov and Sergei Korolev wrote their historic proposal to the Soviet government and party, *Most promising works in the development of outer space*. Among other things, they proposed:

- The landing of small 10 kg to 20 kg research stations on the moon.
- A satellite to photograph the lunar surface.
- A lunar flyby, with the subsequent recovery of the payload to Earth.

Noting the American attempts to orbit the moon with Pioneer, Korolev made a proposal to government in February 1959 for a small probe to orbit the moon, the Ye-5. However, this required a heavier launcher than was available; and, in any case, the proposal was subordinated to the need to achieve success with the

Ye-1 to -4 series, which was proving difficult enough. The Ye-5 never got far. Korolev revised his proposals in late 1959, by which time a much more advanced upper stage was now in prospect, one able to send 1.5 tonnes to the moon, a considerable advance, but a figure identified by Tikhonravov as far back as 1954 in *Report on an artificial satellite of the Earth*. By now, the proposal was for:

- A new lunar rocket and upper stage, the 8K78, later to be called the Molniya.
- A lander, called the Ye-6.
- An orbiter, the Ye-7.

These were approved by government during the winter of 1959–60. OKB-1 Department #9, under Mikhail Tikhonravov, was assigned the work and he supervised teams led by Gleb Maksimov and Boris Chertok. Design and development work got under way in 1960, but it does not seem to have been a priority, the manned space programme taking precedence. The 8K78 was primarily designed around the payloads required for the first missions to Mars and Venus, rather than the moon, but they equally served for the second generation of Soviet lunar probes.

NEW LUNAR ROCKET

The new rocket, the 8K78, was a key development. The 8K78 became a cornerstone of the Soviet space programme as a whole, not just the moon programme and versions were still flying over 40 years later, over 220 being flown. The following were the key elements:

- Improvements to the RD-108 block A and RD-107 block BVGD stages of the R-7, with more thrust, higher rates of pressurization and larger tanks, developed by Glushko's OKB-453.
- A new upper stage, the block I, developed with Kosberg's OKB-154.
- A new fourth stage, the block L, designed within OKB-1.
- New guidance and control systems, the I-100 and BOZ.

In a new approach, the first three stages would put the block L and payload in Earth orbit. Block L would circle the Earth once in what was called a parking orbit before firing out of Earth orbit for the moon. With the Ye-1 to -4 series, a direct ascent was used, the rocket firing directly to the moon. The problem with direct ascent was that even the smallest error in the launch trajectory, even from early on, would be magnified later. By contrast, parking orbit would give greater flexibility in when and how rockets could be sent to the moon. The course could be recalculated and readjusted once in Earth orbit before the command was given. Parking orbit also enabled a much heavier payload to be carried.

The principal disadvantage – no one realized how big it would turn out to be – was that the engine firing out of parking orbit required the ignition of engines that had been circling the Earth in a state of weightlessness for over an hour. This was where

The 8K78

block L came in. Block L was designed to work only in a vacuum, coast in parking orbit and then fire moonward. A device called the BOZ (*Blok Obespecheyna Zapushka*) or Ignition Insurance System would guide the firing system toward the moon. Block L was 7.145 m long, the first Soviet rocket with a closed-stage thermodynamic cycle, with gimbal engines for pitch and yaw and two vernier engines for roll. The new third stage, block I, was based on an intercontinental ballistic missile design called the R-9. A new orientation system for blocks I and L, called the I-100, was devised by Scientific Research Institute NII-885 of Nikolai Pilyugin.

8K78 Molniya rocket

Total length	44 m
Diameter (blocks BVGD)	10.3 m
Total weight	305 tonnes
of which, frame	26.8 tonnes
propellant	279 tonnes
Burn time first stage (block A)	301 sec
Burn time second stage (blocks BVGD)	118 sec
Burn time third stage (block I)	540 sec
Burn time fourth stage (block L)	63 sec

The new 8K78 rocket, including block L, was built in some haste. Block L was ordered in January 1960 and the blueprints approved in May. The first two stages, with block I but without block L, were fired in suborbital missions from January onward. Block L was first tested aboard Tupolev 104 aircraft, designed to simulate weightlessness, in summer 1960. The first all-up launchings took place in October 1960, when two probes were fired to Mars, both failing at launch. Two Venus launches were made in February 1961, one being stranded in Earth orbit but the second one getting away successfully. But the worst period in the development phase was still to come. Three Venus probes in a row failed in August/September 1962, all at launch. Of three Mars probes in October/November 1962, only one left parking orbit. Blocks A and B failed once, block I three times and block L four times. The Americans later published the list of all these failures (this took the form of a letter to the secretary general of the United Nations from ambassador Adlai Stevenson on 6th June 1963), but some people assumed they were making them up, for no country could afford so many failures and still keep on trying.

YE-6 LUNAR LANDER

The lunar lander was called the Ye-6. In the event, there were two variants: the Ye-6, used up to the end of 1965; and the Ye-6M, used in 1966. The Ye-6 series had two modules. The main and largest part, the instrument compartment, was cylinder-shaped, carried a combined manoeuvring engine and retrorocket, orientation devices, transmitters and fuel. The lander, attached in a sphere on the top, was quite small, only 100 kg. It was ball-shaped and once it settled on the moon's surface, a camera would peep up to take pictures. It followed very closely the popular image of what an alien probe landing on Earth would look like.

The main spacecraft was designed to carry the probe out to the moon and land it intact on the surface. The engine, built by Alexei Isayev's OKB-2, would be fired twice: first, for a mid-course correction, with a maximum thrust of 130 m/sec; and, second, to brake the final stage of the descent. The engine was called the KTDU-5, an abbreviation from *Korrektiruiushaya Tormoznaya Dvigatelnaya Ustanovka*, or course correction and braking engine) and it ran off amine as fuel and nitric acid as oxidizer. The next most important element was the I-100 control system, built by Nikolai Pilyugin's Scientific Research Institute NII-885. This had to orientate the spacecraft properly for the mid-course correction and the landing. The mid-course correction was intended to provide an accuracy of 150 km in the landing site. The main module relied on batteries rather than solar power.

The final approach to landing would be the most difficult phase. The rocket on the 1,500 kg vehicle had to fire at the correct angle about 46 sec before the predicted landing. It must brake the speed of the spacecraft from 2,630 m/sec 75 km above the moon to close to 0 during this period. Too early and it would run out of fuel before reaching the surface, pick up speed again and crash to pieces. Too late and it would impact too fast. The main engine was designed to cut out at a height of 250 m. At this stage, four thrusters were expected to slow the spacecraft down to 4 m above the

The Ye-6 lander

surface. A boom on the spacecraft would then detect the surface. As it did so, gas jets would fill two airbags and the lander would be ejected free to land safely. Four minutes after landing, a timer would deflate the bags and the lander would open from its shell.

Ye-6 Luna
Height	2.7 m
Base	1.5 m approx.
Weight	1,420 kg
Engine thrust	4,500 kg

Landing cabin

Height	112 cm
Base	58 cm
with petals	160 cm
with arms	3 m
Weight	82 kg

Ye-6 instruments
- Ye-6M (Luna 13).
- Camera.
- Radiometer.
- Dynamograph/penetrometer ('gruntmeter').
- Thermometer.
- Cosmic ray detector.

The lander was egg-shaped, pressurized, metallic-looking and made of aluminium. Inside were a thermal regulation system, chemical batteries designed to last four days, transmitters and scientific equipment. Once stable on the surface, four protective petals would open on the top to release the four 75 cm transmitting aerials. The most important element was of course the camera. Although often described as a television camera, it was more accurately called a pinpoint photometer and took the form of a cylinder with a space for the scanning mirror to look out the side. These are optical mechanical cameras and do not use film in the normal sense, instead scanning for light levels, returning the different levels by signal to Earth in a video, analogue or digital manner. The system was designed by I.A. Rosselevich, built by Leningrad's Scientific Research Institute NII-380 and was based on systems originally used on high-altitude rockets. The camera was small, only 3.6 kg in weight and used a system of mirrors to scan the lunar surface vertically and horizontally over the period of an hour working on only 15 watts of electricity. The lander would transmit for a total of five hours over the succeeding four days, either on pre-programmed command or on radioed instructions from the ground.

A safe landing required as vertical a descent as possible. From the photography point of view, the Russians wanted to land a spacecraft during local early dawn. The lunar shadows would therefore be as long as possible, providing maximum contrast and enabling scale to be calculated. Once again, Keldysh's Mathematics Institute calculated the trajectories. Earth–moon mechanics and lighting conditions were such that a direct early dawn descent could come down in only one part of the moon, the Ocean of Storms. This is the largest sea on the moon, covering much of its western hemisphere.

The Americans built a comparable spacecraft, Ranger. Here, the Americans intended to achieve the double objective of photographing the lunar surface and achieve a soft-landing. On Ranger, the main spacecraft was a hexagonal frame which contained the equipment, engine and cameras. As Ranger came down toward the lunar surface, photographs would be taken until the moment of impact. Ranger's soft-landing capsule would use a different landing technique: 8 sec before impact and at an altitude of 21.4 km, the landing capsule, with a retrorocket, would separate from the crashing mother craft. The powerful solid rocket motor would cut its speed. The cabin would separate, impact at a speed of not more than 200 km/hr and then bounce onto the lunar surface. Ranger's landing capsule was about half the size and weight of the Ye-6. It was made out of balsa wood and the instruments would be protected by oil. There was a transmitter and only one instrument: a seismometer (no camera).

THE TRACKING SYSTEM

A tracking station had already been built for the moon probes of 1958–60, located in the Crimea. Its southerly location was best for following a rising moon. The Crimea around Yevpatoria offered several advantages for a tracking system. Originally, the tsars had built their summer homes around there and it had now become a resort area, meaning that it was well served by airfields. There were defence facilities in the region and military forces who could assist in construction.

The tracking system was considerably expanded in 1960. This was done to serve the upcoming programme for interplanetary exploration, but these new facilities could also be used for lunar tracking. The new construction at the Yevpatoria site was called the TsDUC, or Centre for Long Range Space Communications. The TsDUC actually comprised two stations with two receivers (downlink) and one transmitter (uplink), facilitated by a microwave station, which transmitted data from the receiver stations to another microwave system in nearby Simferopol and thence on to other locations in the USSR. The records are confusing about what was actually built at the time and where and little was said about them publicly, presumably to hide Soviet tracking capabilities from the snooping Americans. We know that the Americans had good intelligence maps of the Yevpatoria system from 1962, but it would be surprising if they had not had good details a little earlier.

For the moment, two sets of eight individual duralium receiving dishes of 15.8 m were built on a movable structure, designed to tilt and turn in unison. Two were built 600 m apart at what the Americans called 'North Station' and a set of half the size, 8 m transmitting dishes called Pluton at what they called 'South Station'. North Station was the largest complex of the two, surrounded by 27 support buildings, 15 km west of Yevpatoria. To construct the receiving stations, Korolev was forced to improvise. He came up with the idea of using old naval parts for the station: a revolving turret from an old battleship, a railway bridge for support and the hull of a scrapped submarine. They received signals on the following frequencies: 183.6, 922.763, 928.429 MHz and 3.7 GHz.

South Station was to the southeast and much closer to Yevpatoria, 9 km. It comprised one, later eight 8 m dishes in a similar configuration to, but half the size of the duo at North Station. Transmission power was rated at 120 kW and its range was estimated at 300 million km. Transmissions were sent at 768.6 MHz.

Dishes at Yevpatoria

Even though chief designer Yevgeni Gubensko died in the middle of construction, Yevpatoria station went on line on 26th September 1960, just in time for the first, but unlucky Mars probes. The facilities there were originally quite primitive, ground controllers being provided with classroom-style desks, surrounded by walls of computer equipment. Modern wall displays did not come in until the mid-1970s. Still, it was the most powerful deep space communications system until NASA's Goldstone Dish came on line in 1966. In 1963, just in time for the new Ye-6 missions, the lunar programme acquired a dedicated station, a 32 m dish in Simferopol called the TNA-400.

Until a mission control was opened in Moscow in 1974, Yevpatoria remained the main control for all Russian spaceflights, not just the interplanetary ones. It was normal for the designers to fly from Baikonour Cosmodrome straight to Yevpatoria to oversee missions. The Americans, by contrast, had a worldwide network of tracking stations, with large dishes in California, South Africa and Australia. Dependence on one station at Yevpatoria imposed two important limitations on Soviet lunar probes. First, the arrival of a spacecraft at the moon had to be scheduled for a time of day when the moon was over the horizon and visible in Yevpatoria, so schedules had to be calculated with some care in advance. Second, as noted during the 1959 missions, there was no point in having Soviet moon probes transmit continuously, for their signals could not be picked up whenever the moon was out of view. Instead, there would be short periods of concentrated transmission, called 'communications sessions' scheduled in advance for periods when the probes would be in line of sight with Yevpatoria. This required the use of timers and sophisticated systems of control, orientation and signalling.

Korolev and his colleagues attempted to get around the limits imposed by the Yevpatoria station. If they lacked friends and allies abroad to locate tracking dishes, there were always the oceans. Here, three merchant ships were converted to provide tracking for the first Mars and Venus missions, but they could also serve the moon programme. These ships were the *Illchevsk*, *Krasnodar* and *Dolinsk* and their main role was to track the all-important blast out of parking orbit, which was expected to take place over the South Atlantic. The ships were a helpful addition, but they had limitations in turn. First, ships could not carry dishes as large as the land-based dishes; and, second, they were liable to be disrupted in the event of bad weather at sea, which made it difficult to keep a lock on a spacecraft in a rolling sea.

LUNA 4 AND THE 1963, 1964 ROUNDS OF LAUNCHINGS

Throughout 1962, the Ye-6 was put through a rigorous series of ground tests. These focused on the landing sequences, the operation of the airbags and ensuring their subsequent successful deployment.

The first Ye-6 was successfully launched into Earth parking orbit on 4th January 1963, four years and two days after the First Cosmic Ship. Block L was due to fire from its parking orbit over the Gulf of Guinea toward the end of the first orbit to send the new spaceship moonbound. The *Dolinsk* was steaming below to track the signals.

Once again, the block L let everyone down. The power system in the I-100 control unit appears to have failed, for the electrical command to ignite block L was never sent. The moon probe orbited the Earth for a day before breaking into fragments and burning up. A second attempt was made a month later, on 3rd February. Control of the pitch angle began to fail at 105.5 sec. I-100 control was lost just as block I was due to fire. There was no third-stage ignition and the two upper stages crashed into the Pacific near Midway Island. Both launches were detected by the Americans, who had no difficulty in assessing them as failed moon probes.

Sorting out the I-100 control unit took two months. The next probe was launched on 2nd April 1963 and became the first Russian moon probe to leave a parking orbit for the moon. It was named Luna 4 (no more 'cosmic ships' or 'interplanetary stations'), although in reality it was the twelfth Russian moonshot. Its precise purpose was not revealed, except to say that it would travel to 'the vicinity of the moon'. Although the Russians did not specifically ask Jodrell Bank to track Luna 4, they issued transmission frequencies (183.6 MHz) and gave navigational data, an indirect invitation to do so. Jodrell Bank picked up signals for six hours, two days after the probe left Earth. The Russian receiving stations followed the mission from their new base in the Crimea and the spacecraft was also picked up visually as a 14th magnitude star. The Soviet news agency, Tass, was upbeat:

Scientists have to clarify the physical conditions cosmonauts will meet, how they are to overcome landing difficulties and how they should prepare for a prolonged stay on the moon. The human epoch in the moon's history is beginning. There will be laboratories, sanatoria and observatories on the moon.

This heady enthusiasm soon evaporated. The following day, it became clear that the astro-navigation system had failed and that it would be impossible to perform a mid-course manoeuvre. The next day, on 4th April, the USSR reluctantly announced that Luna 4 would fly 'close to' the moon at 9,301 km the following day (in reality, it may have come slightly closer, 8,451 km). Jodrell Bank listened in carefully for 44 min during the point of closest passage. Contact was lost two days later and Luna 4 ended up in a highly eccentric equatorial Earth orbit of 89,250 by 694,000 km, taking 29 days per revolution and may have been eventually perturbed out of it into solar orbit. The Russians claimed – quite unconvincingly – that a lunar flyby was all that had been intended. But they shut up about health resorts on the moon for the time being.

The three failures in four months forced a review of the programme, this time headed up by Mstislav Keldysh, who was now president of the Academy of Sciences. The investigators never determined the true cause of the failure of Luna 4. All that was known for certain was that the mid-course correction had never taken place because the astro-navigation system had failed, which meant that the spacecraft could not be orientated for the burn in the first place. The Keldysh investigation did find many problems with the system itself and these were corrected over the following year. There was abundant evidence of the programme being prepared in too much of a hurry and quality control suffering as a result.

It was another year before the next Ye-6 was made ready for launch. The background was not propitious, for two more 8K78 Molniya rockets with test probes for Venus had failed in the past six months. What should have been Luna 5 was launched on 21st March 1964, but a rod broke in the block I stage, a valve failed to open fully, it never reached full thrust, cut off at 489 sec and the stage crashed back to Earth. On the 20th April 1964, a month later, the next Ye-6 suffered the same fate, but this time the connecting circuitry between the BOZ and the I-100 failed, the mission ending after 340 sec. Despite further efforts to resolve the problems in the upper stage, the next moon rocket was lost as well on 12th March 1965. This time, block L failed to ignite due to a transformer failure. The mission was given the designator of Cosmos 60, but the ever-watchful Americans knew at once that it was a moon failure. Confirmation that this was the case came when, many years later, it became known that Cosmos 60 had carried a gamma ray detector of the type later flown on Luna 10 and 12. Even though the mission failed as a moon probe, useful scientific results on cosmic rays were obtained [1].

This time, more significant steps were taken to address the problems of integrating block 1, block L, the BOZ and the I-100. The whole system was re-worked and re-wired, with separate control systems installed on both block L and the Ye-6. Little good did it do, for the next Luna crashed to destruction on 10th April 1965. This time the pressurization system for the liquid oxygen tank of block I failed, causing the spacecraft to crash into the Pacific. The new guidance system was never tested. This was the fourth failure in a row since Luna 4. Indeed, since the Automatic Inter-planetary Station, Russia had attempted to launch nine probes to the moon, none had been successful and only one had been announced. The level of failures represented a rate of attrition no programme could sustain and questions were being asked in the Kremlin by now.

INTRODUCING GEORGI BABAKIN

These setbacks led to a major shakeup in the moon programme. Korolev's OKB-1 was now heavily overcommitted and the manned space programme was using up his full energies. Korolev approached the Lavochkin Design Bureau. This was, at first sight, a strange thing to do, for Lavochkin was an aircraft design bureau that had languished since the death of its founder, Semyon Lavochkin. This design bureau dated to 1937, being founded as Plant #301 by aviation designer Semyon Lavochkin. During the 1940s the plant made fighter aircraft and during the 1950s, cruise missiles. Plant #301 was named the Lavochkin Design Bureau on the death of its founder in 1960. The deputy director then was Georgi Babakin but he had since gone to work for Korolev's rival, Vladimir Chelomei.

Georgi Babakin is to become a central person in our story. Fifty-year-old Georgi Babakin was an unusual man, self-taught, with a healthy suspicion of formal educa-tion. Born in Moscow on 31st August 1914 (os), he developed an early passion for radio electronics, becoming senior radio technician with the Moscow Telephone Company in 1931. He was drafted into the Red Army's Proletarian Infantry Division in 1936 where he was radio operator for six months before being dismissed for ill

Georgi Babakin

health. He returned to school, where he completed his exams, joining the old Lavoch-kin Design Bureau during its plane-making days, rising to deputy chief designer. He eventually took a university degree in 1957 [2].

March 1965 saw a shakeup in the unmanned lunar programme in which the Ye-6 missions, as well as the interplanetary programme, left OKB-1. OKB-301 was effec-tively reconstructed, with its former deputy director Georgi Babakin returning as chief designer. Specifically, Korolev asked Georgi Babakin to ask him to take over the Ye-6 programme once the current OKB-1 production run was complete, but he knew that this would mean the entire set of programmes going to Lavochkin from then on. In April 1965, Sergei Korolev made his first and only visit to the Lavochkin Design Bureau. He met all the senior design staff, formally handed over the OKB-1 blueprints to them, made clear the heavy duty now incumbent upon them and warned them that he would take the projects back if they did not perform. Lavochkin's experience of producing military aircraft stood to its advantage, for the company put much emphasis into ground testing and cleaning bugs out of the system beforehand.

Few people seem to have moved across from OKB-1 to Lavochkin. One who did was Oleg Ivanovsky. Another radio enthusiast, he was a cossack cavalryman during the war but was so badly wounded that at war's end he was registered permanently disabled, facing a grim future without work or, more importantly, worker ration cards. An old friend managed to get him work in OKB-1 where his radio skills were quickly appreciated. Korolev gave him a key role in the radio instrumentation for Sputnik, the 1959 moon probes and then the Vostok, personally accompanying Yuri Gagarin to his cabin. When the new Lavochkin company was set up, Korolev found him a post as deputy chief designer, second only to Babakin [3].

At the same time, the Isayev bureau also improved the KTDU-5 engine system. A new version, called the KTDU-5A, was introduced. Using amine as fuel and nitric acid as oxidizer, it had a specific impulse of 278 sec, a thrust of 4,640 kg and a chamber pressure of 64 atmospheres. It was designed to burn twice – the first time for the mid-course correction (up to 130 m/sec) and then a second time for the landing (2,630 m/sec) and had a total burn time of 43 sec [4]. The decision was also taken to upgrade the launcher and replace the unreliable 8K78 and block L by an improved version. The lower stages, the 8K78, were replaced by the 8K78M by the end of the year and the old block L by the new block MVL by 1968.

RETURN TO THE MOON

To what must have been enormous relief in OKB-301, the next moon probe sailed smoothly away from Earth orbit on 9th May 1965. This date marked Victory in Europe Day, 20 years from the end of the war and hopefully this would augur well for the new probe, Luna 5. Maybe the guidance systems had at last been corrected. Nine communications sessions took place *en route* to the moon. During the first five, the probe radioed back its exact position as accurately as possible so that the thrust for the mid-course correction could be calculated. The fifth session issued the commands. Things began to go wrong now. The I-100 was unable to control the probe properly and it began spinning. Ground control brought it back under control and tried again. The command instructions were issued wrongly, so the burn did not take place. By now it was too late to carry out the burn. Thankfully, Luna 5's original path was sufficiently accurate to hit the moon, although far from the area intended, so an embarrassing repeat of the Luna 4 could be avoided. Ground control positioned the spacecraft for retrofire, aware that the spacecraft would come down about 700 km off course and that it would not be the intended direct, vertical descent but an oblique one instead. The I-100 again failed to stabilize the probe, so retrofire did not take place. Soviet scientists in the control room listened helplessly to Luna 5's signals as it crashed unaided on the moon at great speed, way off course. Its precise impact point has never been determined and the original Soviet announcement suggested the Sea of Clouds, a location of 30°S, 8°W being later suggested. Some subsequent analysis gave an impact point to the northwest and nearer the equator (8°10′N, 23°26′W), but well away from the Sea of Clouds [5].

Luna 5 exploded and sent up a cloud of dust measuring 80 km wide and 225 km long. It was the second Soviet probe to impact on the moon, the first since the Second Cosmic Ship seven years earlier. The announcement of the unhappy outcome was not made until twelve hours later: whether this was in the forlorn hope that the probe might have survived, or to give time to put news management into operation, is not known.

The idea that Luna 5 had created a big impact cloud was ridiculed at the time and subsequently. The cloud was seen by observers at Rodewitsch Observatory in the German Democratic Republic until ten minutes after impact when it faded and the details given in *Izvestia* on 16th May. The claims were treated nowhere more seriously than in the United States, where Bellcomm Inc. was commissioned by NASA to investigate. Bellcomm's report was done by J.S. Dohnanyi, who concluded that August that if Luna 5 impacted into a basalt surface and if the fuel of the landing rocket exploded on impact, then such a cloud was indeed possible [6].

Luna mid-course correction

Luna 6 on the 8th June set off for the moon with the same promise as Luna 5. There was a sense of apprehension as the mid-course manoeuvre approached. Although the rocket switched on correctly, it would not turn off! The engine continued to blast away remorselessly, sending Luna 6 away in the opposite direction. It missed the moon by no fewer than 160,935 km, what must have been a record. Trying to salvage something from another disappointment, ground control commanded a separation of the lander and inflation of the airbags, a manoeuvre that apparently worked.

LUNAR FARSIDE PHOTOGRAPHY

After all these Luna disappointments, it was ironic that during the summer the Soviet Union now achieved an unexpected success courtesy of an unlaunched Mars probe. This was Zond 3. The title 'Zond' had been contrived by Korolev to test out the technologies involved in deep space missions. Zond 1 had been sent to Venus in March 1964, while Zond 2 headed for Mars in November 1964, coming quite close to hitting the planet the following summer. These Zonds each had two modules: a pressurized orbital section, 1.1 m in diameter, with 4 m wide solar panels, telemetry systems, 2 m transmission dish, a KDU-414 engine for mid-course manoeuvre and a planetary module. This could be a lander (e.g., Zond 1), but in the case of Zond 3 this was a photographic system, accompanied by other scientific instruments. The probe was compact and smaller than the Lunas at 950 kg. The camera system was a new one introduced for the 1964–5 series of Mars and Venus probes. The designer was Arnold Selivanov and his system was comparatively miniscule, weighing only 6.5 kg. The film used was 25.4 mm, able to hold 40 images and could be scanned at either 550 or 1,100

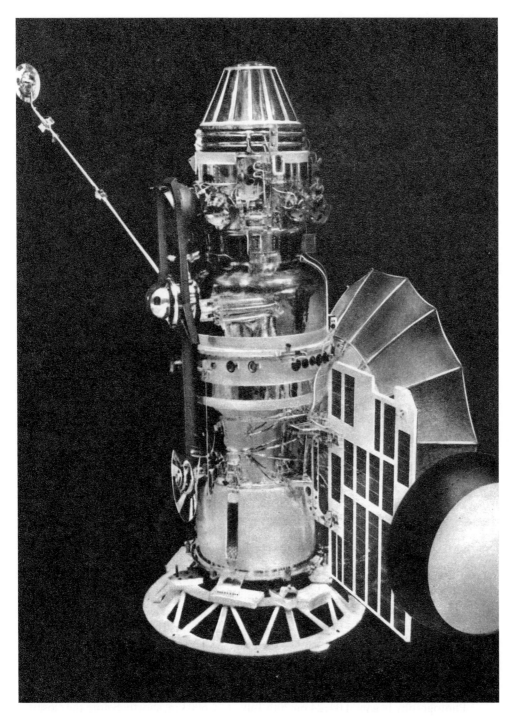

Zond 3

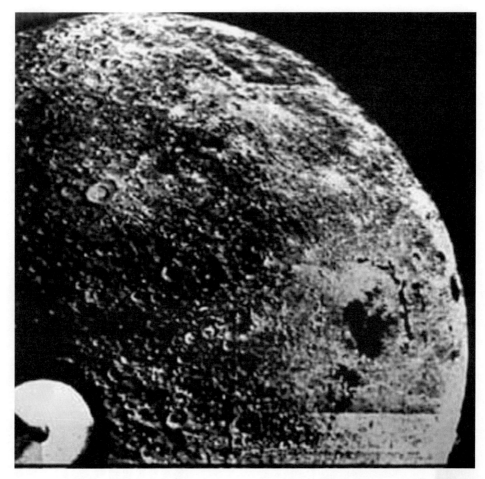

Zond 3 over Mare Orientale

lines. Transmission could be relayed at 67 lines a second, taking only a few minutes per picture, or at high resolution, taking 34 min a picture. Additional infrared and ultraviolet filters were installed.

Zond 3 was supposed to have been launched as a photographic mission to Mars in November 1964 as well, but it had missed its launching window. Now this interplanetary probe was reused to take pictures of the moon's farside and get pictures far superior to those taken by the Automatic Interplanetary Station in 1959 and of the 30% part of the lunar farside covered neither then nor by the April 1960 failures. Taking off on 18th July 1965, nothing further was heard of it until 15th August when a new space success was revealed. Zond 3 had shot past the moon at a distance of 9,219 km some 33 hours after launch *en route* to a deep space trajectory.

Photography began at 04:24 on 20th July at 11,600 km, shortly before the closest passage over the Mare Orientale on the western part of the visible side. Well-known

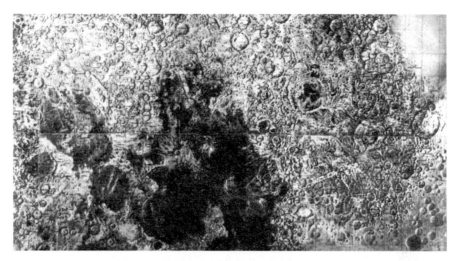

Lunar map after Zond 3

features of the western side of the moon were used to calibrate the subsequent features and the idea was to cover those parts of the moon not seen by the Automatic Interplanetary Station, which had swung round over the eastern limb of the moon. As Zond 3 soared over the far northwestern hemisphere of the moon, its f106-mm camera blinked away for 68 min at 1/100th and 1/300th of a second. By 05:32, when imaging was concluded, 25 wide-view pictures were taken, some covering territory as large as 5 million km^2 and, in addition, three ultraviolet scans were made. The details shown were excellent and were on 1,100 lines (the American Ranger cameras of the same time were half that).

Soviet scientists waited till Zond 3 was 1.25 million km away before commanding the signals to be transmitted by remote control. They were rebroadcast several times, the last photo-relay being on 23rd October at a distance of 30 million km. There was grandeur in the photographs as Zond swung around the moon's leading edge – whole new mountain ranges, continents and hundreds of craters swept into view. Transmissions were received from a distance of 153.4 million km, the last being on 3rd March 1966. Course corrections were made using a new system of combined solar and stellar orientation.

Zond 3 had been built by OKB-1 entirely in-house, not using the I-100 control system. It was the last deep space probe designed within OKB-1, before the moon programme was handed over to Lavochkin.

With Zond 3, the primitive moon maps of the lunar farside issued after the journey of the Automatic Interplanetary Station could now be updated. Whereas the nearside was dominated by seas (*maria*), mountain ranges and large craters, the farside was a vast continent with hardly any *maria*, but pockmarked with small craters. The Russians again exercised discoverers' prerogative to name the new features in their own language. Thus, there were new gulfs, the *Bolshoi Romb* and the *Maly Romb* (big and small) and new ribbon maria *Peny*, *Voln* and *Zmei* [7].

Zond 3: scientific instruments
Two cameras.
Infrared and ultraviolet spectrometer.
Magnetometer.
Cosmic ray detector.
Solar particle detector.
Meteoroid detector.

Zond 3 may have encouraged the designers to believe that in their next soft-landing mission, Luna 7, they would at last meet with success. Launch was set for 4th September 1965, but faults were found in the R-7 control system and the entire rocket had to be taken back into the hangar for repairs, missing the launch window. Luna 7 left Earth the following month, on the eighth anniversary of Sputnik's launch, on 4th October. On the second day, the mid-course correction burn went perfectly, unlike what had been the case with Luna 5 or 6. On the third day, two hours before landing and 8,500 km out, the Luna 7 orientated itself for landing. Unlike Luna 5, it was on course for its intended landing area near the crater Kepler in the Ocean of Storms. As it did so, the sensors lost their lock on the Earth and, without a confirmed sensor lock, the engine could not fire. This was the second time, after Luna 4, that the astro-navigation system had failed. Ground controllers watched helplessly as Luna 7 crashed at great speed, much as Luna 5 had done only months earlier. Investigation found that the sensor had been set at the wrong angle, in such a way that it would find it difficult to locate and hold Earthlock in the first place.

Korolev was summoned to Moscow to explain the continued high failure rate. His old patron, Nikita Khrushchev, had now been deposed and Korolev now had to deal with the new leadership around Leonid Brezhnev. Korolev admitted that there had been great difficulties and promised success the next time. Luna 8 was duly launched on 3rd December. This was the last of the Ye-6 production run of OKB-1. Luna 8 used a new parking orbit. Its predecessors, Luna 4–7, has used a parking orbit of 65° to the equator. Now, a lower equatorial angle of 51.6° was used, making it possible to increase the mass of the spacecraft from around 1,500 kg to around 1,600 kg.

Luna 8 smoothly passed the hurdle of the mid-course correction. This time it got into a correct position for the deceleration burn and a descent to crater Kepler. Now, at this late stage, things began to go wrong. When the command was sent to inflate the airbags, a sharp bracket pierced one of them and the escaping air set the probe spinning. This blocked the system from orientating itself and the engine from firing. The probe briefly came back into position and the engine fired for 9 sec, before going out of alignment again and cutting out. A 9 sec firing instead of 46 sec clearly did little to prevent what must have been another explosive impact. The decision was taken for the future to inflate the airbags only at the very end of the deceleration burn. This was the tenth failure to achieve a soft-landing.

KOROLEV DIES: THE MISHIN SUCCESSION

Korolev was summoned to Moscow to explain why the promised success had not been forthcoming, but the meeting never took place. He was dead. He was admitted to hospital on 13th January 1966, for the removal of a colon tumour. No less a person than the Minister for Health, Dr Boris Petrovsky, carried out the operation – on Korolev's own request. Mid-way through, Petrovsky discovered a more serious tumour, 'the size of a fist'. He continued the operation. A large blood vessel burst; haemorrhaging began; and Sergei Korolev's heart – weakened as it had been from the toil of the labour camps – collapsed. Attempts to ventilate him were made more difficult by his jaw having been broken by a camp guard during the Gulag years. Frantic efforts were made to revive him, but on 14th January he was pronounced dead.

Once dead, his identity and importance could safely be revealed and indeed it was, following burial in the wall of the Kremlin on 16th January 1966. A flood of Korolev literature followed. No efforts were spared telling of his boundless energy, iron will, limitless imagination and engineering genius. This could have been mistaken for nostalgia but it was not. With Korolev's death, the Soviet space programme was never the same again. The driving force went out of it and with him that unique ability to command, inspire, bargain, lead, design and attend to detail. After 1966, the programme had many excellent designers, planners, politicians, administrators and prophets, but never in one person all together. Not that this was immediately obvious. The programme continued on much as before. But the sense of direction slackened. Indeed, the absence of Korolev may have made the critical difference to the climax of the moon race in 1968–9.

The succession was not clear and the defence minister Dmitri Ustinov proposed Georgi Tyulin who for several months appeared to be the likely new chief designer. In May, the choice eventually fell on Korolev's deputy, Vasili Mishin, who had worked alongside him since 1945. Vasili Mishin – born 5th January 1917 (os) – came from Orekhovzvevo near Moscow and became a mathematician at the Moscow Aviation Institute. Mishin had been the youngest member of Tikhonravov's group to visit Poland in 1944 and had probably done the most to extract what could be learned from the fragments recovered. He was a very bright young engineer and was also a successful pilot. Mishin contributed to the design of Sputnik before being named deputy to Korolev in 1959. He invented, for example, the railcar system for erecting the R-7 on its pad, one which facilitated launches in rapid succession at the same pad and would have enabled the assembly of the Soyuz complex. Vasili Mishin was a kindly man, well regarded by those who interviewed him and, before his death in 2001, did much to tell us of the moon race and open the historical record. Khrushchev made this judgement of him and, while it is harsh, few would dispute it:

Vasili Mishin was excellent at calculating trajectories, but did not have the slightest idea how to cope with the many thousands of people, the management of whom had been loaded onto his shoulders, nor to make the huge irreversible government machine work for him [8].

Cosmonaut Alexei Leonov described him as a good engineer, but hesitant, un-inspiring, poor at making decisions, over-reluctant to take risks and bad at managing the cosmonaut corps [9]. He had a drink problem, though Alexei Leonov observed from first hand that his engineering judgement was remarkably unaffected while still under the influence. OKB-1 was reorganized and renamed TsKBEM (Central Design Bureau of Experimental Machine Building) while Chelomei's bureau was renamed TsKBM (Central Design Bureau of Machine Building) (to avoid obvious and needless confusion, the old designators will continue to be used in this narrative).

The chief designer system had worked well for the Soviet Union in the time of Korolev. But the system was extremely dependant on one person and, lacking Korolev's strengths and skills, the system exposed serious weaknesses when dependent on Mishin. The rival American programme was never as dependent on personality as was the Soviet system. Although Wernher von Braun was the closest the Americans came to a 'great designer', the Americans were much more circumspect in separating the space programme's administrative leadership – the *administrator* of NASA, note the title – from its engineering leadership (the NASA centres and the contracting companies).

The Ye-6 series, its OKB-1 production run now expended, gave way to the Ye-6M series. This was the first series actually built by Lavochkin. The improvements of the Ye-6M might have happened anyway, but were also prompted by the failures of the Ye-6. These were:

- Inflation of the airbags after ignition of the final rocket engine firing.
- New, lighter and more efficient camera system.
- More instruments: two folded booms to be fitted to later spacecraft.

The new cabin was slightly heavier, up from 82 kg to 100 kg. The camera system, designed by Arnold Selivanov and built by NII-885, weighed 1.5 kg, used only 2.5 watts of power, could see a horizon 1,500 m distant and was in the form of a rotating turret out of the top of the lander. It was designed to have a higher resolution than the cameras on Luna 4–8 and a full 360° panorama would have 6,000 lines.

'A MAJOR STEP TOWARD A MANNED LANDING'

Lavochkin's Luna 9, 1,583 kg, got away on 31st January 1966. The next day, a course correction manoeuvre was carried out, with 233,000 km still to go. Two days later, 8,300 km out, Luna 9 assumed a vertical position in the line of direction of the moon. The retrorockets blazed into life 75 km above the lunar surface for 48 sec. They cut out when the craft was a mere 250 m above the surface, its downward velocity halted. The lander, cocooned in its two airbags, was flung free to bounce over the lunar surface. The little 99.8 kg capsule settled. On the top part, four petal-like wings unfolded. Four minutes later, four 75 cm aerials poked their way out of the dome, slowly extending. Transmitters working on 183 MHz were set both on the antenna and into the petals.

Arnold Selivanov

Luna 9 was down in the Ocean of Storms between the craters of Reiner and Maria, at 64.37°, 7.08°N. On Earth, ground controllers gathered that winter night, in a moment of expectation. The main module's signals had died as it crashed. Ground controllers had to wait a very long 4 min 10 sec to know if the capsule had survived, if its instruments were still functioning and whether they would get usable information. Or would they be robbed of victory yet again?

Luna 9 cabin on the moon's surface

Luna 9 camera system

They were not. It was an agonizing wait. Exactly on schedule, from the walls of the control room, where loudspeakers had been installed, flooded in the beeps, pips and humming squeal of the signals of Luna 9, direct from the surface of the moon! It was a sweet moment. For the first time a spacecraft was transmitting directly to Earth from another world! It was a historic moment and Radio Moscow lost no time in assessing its significance:

A soft-landing was one of the most difficult scientific and technical problems in space research. It's a major step towards a manned landing on the moon and other planets.

Its importance was not lost on the West. 'New space lead for Russia' and 'Russians move ahead again' were typical headlines. Once again, America's equivalent project called Surveyor had managed to get itself a year or two behind schedule. American engineers were quick to point out that their craft was much more sophisticated than Luna 9. Surveyor planned a real parachutist-type landing, with no rough-landing capsules and would do a more detailed job. 'But, so what?,' people asked. It was still in the shed.

For the next few days the eyes of the world focused on the moon. At 4:50 a.m. on 4th February the television camera was switched on, a mere 60 cm above the surface. Using its mirror, it began to scan the lunar surface, a process which took 100 min through the full 360°. The camera was in fact the main instrument on board: the other

Luna 9 first image

was a radiation recorder. A series of communication sessions was held between Earth and the moon over the next three days, which was as long as the batteries permitted. Some sessions lasted over an hour. The principal one was held on 4th February from 6:30 p.m. to 7:55 p.m. Moscow time and it sparked off a diplomatic incident. This relay was the big one – the one with the pictures. Presumably, they would be sent in code and the best would eventually be published by the Soviet media in the fullness of time.

The ever-present radio dish at Jodrell Bank, Manchester had been listening in to the Luna 9 signals all during the flight. The public relations officer at Jodrell Bank had worked in newspapers and he at once recognized the sound of the next round of signals: it was just like a newspaper's fax machine! On an inspiration and a hunch a car was speedily despatched to Manchester to pick up the *Daily Express* fax machine. The car collected it, dashed back and the fax machine was at once linked up to the radio receiver and the signals from the moon. Reporters were breathless as the fax at once began converting the signals into standard newsroom photographs.

The print was passed round to observatory director Sir Bernard Lovell and the animated reporters. They could only gasp. The tall, authoritative Sir Bernard drew a deep breath and could only utter 'amazing!' There it was, at last – the face of the moon. The camera's eye stretched to a sloping horizon and there loomed rocks, pebbles, stones and boulders, scattered randomly across a porous rocky surface. In the far distance was a crater dipping down. Long shadows accentuated the contrasts of the other-worldly glimpses of the moon's stark surface.

Luna 9, panoramas 1, 2, 3

In minutes, the world had seen the photographs. They were put out at the top of the television news and carried by every national newspapers' front page the next day. The Russians were furious at the West scooping 'their' pictures. Sir Bernard was accused of sensationalism and irresponsibility. In fact, he had his scales slightly wrong (he had no means of guessing the right scale) and the real pictures were slightly flatter.

Luna 9's transmissions were to continue for several days. Two further photographic transmissions followed, a second one on the 4th and the third on the 5th February. More photographs were picked up – eight in all – showing the lunar horizon stretching 1.4 km away in the distance and Luna 9 obviously settled into a boulder field. It was eerie, but reassuring. If Luna 9 could soft-land, then so too could a manned spacecraft: it would not suffocate in a field of dust as some had feared. As the cooling water in the spacecraft gradually evaporated, it settled its position, giving a new angle to the photos.

Its battery exhausted, Luna 9 finally went off the air forever on 6th February after transmitting 8 hr 5 min of data over seven communication sessions, of which three were photo-relays. The Ocean of Storms returned to its customary silence and

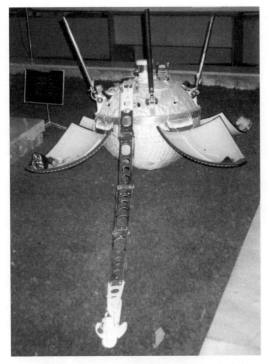

Luna 13

desolation. Luna 9 had done all that was asked of it: it had survived, transmitted and photographed and its timer had ensured regular broadcasts from the moon to the Earth. Another barrier to a lunar landing was down and the Americans had been beaten once again.

The second and final Ye-6, called the Ye-6M, was Luna 13. Luna 13 went aloft on 21st December 1966, fired its retrorockets at a distance of 69 km and bumped down onto the surface of the moon in the Ocean of Storms between the craters Craft and Selenus some 440 km distant from the Luna 9 site at 18.87°N, 62.05°W. For six days, till 28th December, it sent back a series of panoramic sweeps of surrounding craters, stones and rocks. Sun angles were low and the shadows of Luna 13's antennae stood out against the ghostly lunarscape. Five full panoramas were returned.

Luna 13 was more advanced than its soft-landing precursor Luna 9. This was also reflected in its increased weight, 112 kg. Luna 13 carried two extensible arms which folded down like ladders onto the lunar surface. Each was 1.5 m long. At the end of one was a mechanical soil meter with a thumper which tested the density of the soil to a depth of 4.5 cm with an impulse of 7 kg and force of 23.3 kg/m^2. This was developed by Alexander Kemurdzhian at VNII Transmash, the company later responsible for the development of lunar rovers. The conclusion was that moonrock was similar to medium-density Earth soil (800 kg/m^3, to be precise), solid, not dusty. The moon dust was between 20 cm and 30 cm deep. At the end of the other arm was a radiation density

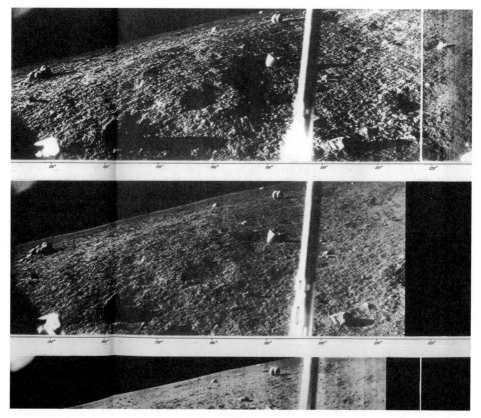

Luna 13 images

meter which found that levels of radiation on the moon were modest and would be tolerable for humans.

Luna 13 also carried a thermometer and a cosmic particle detector. The last found that the moon absorbed about 75% of cosmic ray particles reaching the surface, reflecting the balance, 25%, back into space. The temperature of the lunar surface was measured (117°C). There was little radiation in the lunar soil itself. Table 4.1 shows the key events in the Luna 13 exploration programme.

THE LANDERS: A SUMMARY

Landing a cabin on the moon proved to be much more difficult than either the United States or the Soviet Union imagined. The USSR made twelve attempts, succeeding only twice. With Ranger, the Americans made three attempts, not succeeding until Surveyor 1 in June 1966. For the Russians, the main problems turned out to be the upper stage of the rocket, the translunar course and the navigation systems more than

Table 4.1. Key events in the programme of Luna 13.

No.	Date	Time	Event
1	24 December	21:01	Landing
2	24 December	21:05.30	First signal
3	24 December	21:06–21:18	First communication session
4	25 December	15:15–16:53	Second communication session
5	26 December	16:00–18:23	Third communication session
6	27 December	16:46–19:55	Fourth communication session
7	27 December	20:30–21:32	Fifth communication session
8	27–28 December	23:02–00:21	Sixth communication session
9	28 December	00:41–01:06	Seventh communication session
10	28 December	02:23–02:48	Eighth communication session
11	28 December	07:05–09:13	Ninth and last communication session

the actual landing phase itself. The Ye-6 landers were more sophisticated than the Ranger landers, being able to carry out a broader range of experiments and observations. They achieved the function of determining that the soil would bear a manned spacecraft and that radiation levels on the moon were acceptable. They returned detailed close-up pictures of the moonscape.

The Ye-6 design was later put to good use when the Soviet Union came to soft-land spacecraft on Mars in 1971 (Mars 3). When the American *Pathfinder* successfully soft-landed on Mars in 1997, it used the airbag technique developed by the Russians in the 1960s (not that this was remembered at the time). Airbags were used for the later American Mars probes *Spirit* and *Opportunity* in 2003–4. The difficulties the Americans experienced with their Mars probes also echoed some of the frustrating difficulties experienced by the Russians in the 1960s.

Ye-6, Ye-6M series
4 Jan 1963	Failure
2 Feb 1963	Failure
2 Apr 1963	Luna 4 (missed moon)
21 Mar 1964	Failure
12 Mar 1965	Failure (Cosmos 60, but some science data)
10 Apr 1965	Failure
9 May 1965	Luna 5 (crashed)
8 Jun 1965	Luna 6 (missed moon)
4 Oct 1965	Luna 7 (crashed)
3 Dec 1965	Luna 8 (crashed)
31 Jan 1966	Luna 9 (Ye-6M)
21 Dec 1966	Luna 13 (Ye-6M)

Luna 13 silhouettes

Ye-6, -6M: scientific outcomes

- Density of lunar regolith similar to medium-density Earth rock, little dust, 0.8 gm/cm^3.
- Well able to receive a manned lunar landing vehicle.
- Radiation level of 30 mrad/day, acceptable to humans.
- Moon absorbs three-quarters of cosmic radiation.
- Characterization of local landscape in two locations.
- Temperature of lunar surface, 117°C.

NOW FOR LUNAR ORBIT

Orbiting the moon was as essential to a manned mission as a soft-landing. Good photographs were essential to determine landing sites and it was important to learn as much as possible about the lunar orbit environment to ensure there were no nasty surprises (there were).

The Soviet lunar orbiter programme was commissioned by OKB-1 at the same time as the Ye-6 programme. Called the Ye-7 programme, it made very slow progress in comparison. Two partially completed Ye-7 models were turned over by OKB-1 to OKB Lavochkin in summer 1965 during the move between the design bureaux.

After the success of Luna 9, attention focused on the lunar-orbiting missions.

Luna 10

Although the Ye-7 photographic equipment was not ready, Russia still wanted to achieve a lunar orbit before the Americans did so with their upcoming lunar orbiter. There was also political pressure to mark the 23rd Communist Party Congress, opening at the end of March 1966 and the first congress of new Soviet leader Leonid Brezhnev. Georgi Babakin and Mstislav Keldysh proposed that the Ye-6 bus be used to fly a lunar orbit mission in time for the congress.

This hastily conceived lunar orbiter was called the Ye-6S. It used the Ye-6 bus, to which was attached not the normal lander, but a pressurized 245 kg cabin that would serve as a lunar orbiter. It is more than likely that the cabin was taken from what would have been an Earth-orbiting satellite in the Cosmos series. Its shape strongly suggests that it may have been one of the Cosmos series built by Mikhail Yangel's design bureau in Dnepropetrovsk. It was equipped with seven scientific instruments originally planned for the Ye-7, including a magnetometer on a long boom. From the ground, scientists would also measure gases in the lunar environment by examining signal strengths as the probe appeared and reappeared behind the lunar limb, and watch for changes in the orbit due to the lunar gravitational field. Lunar orbit insertion would be performed by the Ye-6 bus. Instead of a 46 sec burn for soft-landing, a much smaller burn was required for orbit insertion. Once in orbit, the pressurized Cosmos cabin would separate for an independent mission.

The first Ye-6S was launched on 1st March 1966. The upper-stage problems reasserted themselves and block L failed to fire the probe – renamed Cosmos 111 – out of Earth orbit. The second Ye-6S eventually got away on 31st March 1966. No sooner was it streaking towards the moon than it was announced that it was directed towards an entirely new objective – lunar orbit. Eight thousand kilometers from the moon, Luna 10 was turned around in its path and its rockets blazed briefly but effectively. They knocked 0.64 km/sec off its speed, just enough to let it be captured by the moon's gravity field. The boiler-shaped instrument cabin separated on schedule 20 sec later. Luna 10 was pulled into an orbit of 349 by 1,015 km, 71.9°, 2 hr 58 min and became the first spacecraft to orbit the moon.

But, first things first, Luna 10 celebrated the latest Russian achievement in style. Celestial mechanics meant that Luna 10 would enter the first of its lunar orbits just as the Communist Party was assembling in Moscow for its morning congress session. As it rounded the eastern edge of the moon, Luna 10's transmitter went full on and relayed the bars of the *Internationale* – in turn, broadcast live by loudspeaker direct to the party congress over the static of deep space. It was a triumphant moment and the 5,000 delegates had good reason to stand and cheer wildly. Thirty years later, it was learned that the 'live' broadcast was actually a prerecording taken from Luna 10 earlier in the mission. The radio engineers did not trust the live broadcast to work, but, as they later admitted, playing tricks on the Central Committee was a dangerous game and the truth could only be safely revealed in the 1990s when the Central Committee itself was no more.

Luna 10's mission lasted way into the summer and did not end till 30th May after 56 days, 460 lunar revolutions and 219 communication sessions. Data were transmitted on 183 MHz aerials and also on 922 MHz aerials. A stream of data was sent back by its magnetometer, gamma ray spectrometer, infrared radiometer, cosmic ray detector and meteoroid counter. These found a very weak magnetic field around the moon, 0.001% that of Earth (probably a distortion of the interplanetary magnetic field); no lunar magnetic poles; cosmic radiation at 5 particles/cm^2/sec; 198 meteoroid impacts, more in lunar orbit than in the flight to the moon; no gaseous atmosphere; and that there were anomalous zones of mass concentrations below the lunar surface disturbing the lunar orbit (mascons). Using its gamma ray spectrometer, Luna 10 began the first initial survey of the chemistry of the moon, enabling a preliminary map to be compiled. Lunar rocks gave a composition signature broadly similar to basalt, but other important clues to its composition were picked out. The gamma ray spectrometer was used to measure the level of uranium, thorium and potassium in lunar rock. There were significant variations in radiation levels on the moon, being high in

Luna 10 enters lunar orbit

the Sea of Clouds, for example. Luna 10's magnetometer was put on the end of a 1.5 m boom and took measurements every 128 sec for two months. Designer Shmaia Dolginov – who had built the original magnetometer on the First Cosmic Ship – was able to refine the range to between −50 and +50 gammas.

Luna 10's final orbit, as measured on 31st May, was 378–985 km, 72.2° – whether the changes were due to mascons or reflect more accurate measurement of the original orbit is not certain. Despite its hasty assembly, the Dnepropetrovsk Cosmos mission had presented a significant haul of science, significantly advancing the knowledge of the moon in only a couple of months.

Ye-6S

Height	1.5 m
Base	75 cm
Weight (payload)	245 kg
Orbiting altitude	350 × 1,000 km
Plane	71.9°

Luna 10 instruments
Meteorite particle recorder.
Gamma spectrometer.
Magnetometer with three channels.
Solar plasma experiment.
Infrared recorder.
Radiation detector.
Charged particle detector.

Luna 10 cabin

The discoveries of Luna 10
Weak magnetic field around the moon, 0.001%.
No lunar magnetic poles.
Cosmic radiation in lunar orbit.
Meteoroid impacts, more in lunar orbit than in the flight to the moon.
No gaseous atmosphere.
Mascons.
Basaltic surface composition.

THE LUNAR PHOTOGRAPHY MISSIONS

Now that lunar orbit had been achieved ahead of the Americans, the programme could now return to the original, planned Ye-7 lunar-orbiting photography mission. The Ye-7 was renamed the Ye-6LF at this stage. It used the same Ye-6 bus. Instead of the landing cabin, there was a non-detachable cone and box-shaped camera system. Luna 11 carried the same camera system as that flown on Zond 3, which in turn was designed for the 3MV series of Mars and Venus probes over 1964–5. The photographs were expected to cover 25 km^2 each, with a resolution of 15 m to 20 m. Once taken, the photographs would be developed and dried. They would then be scanned by a television system on board. Besides the camera system, seven scientific instruments were carried, the same as the Ye-6S, Luna 10. The whole spacecraft weighed around 1,620 kg.

The first Ye-6LF, with a full photographic suite on board, was eventually launched on 24th August, after the first American lunar orbiter had arrived. Called Luna 11, it left Earth on 24th August and entered moon orbit of 159 by 1,193 km, 27°, 2 hr 58 min. After burning propellant, the mass entering lunar orbit was in the order of 1,136 kg. The Russians had learned their lesson from the Luna 9 episode over the photographs. The Russians faced a choice of sending down pictures only when Yevpatoria was in line of sight, which would take many weeks, or to send them down when stations farther afield, including their own, could pick them up. They decided on the latter course. In a crafty ruse, the decision was taken that transmission would switch rapidly between the two downlink frequencies, too quickly for Jodrell Bank to reconfigure its systems. Moreover, all the photographs were to be taken in the first 24 hours of the mission and transmitted straight away, before this cat-and-mouse technique could be realized or countered.

The Russians reported completion of the mission on 1st October after 38 days, 277 revolutions and 137 communications sessions – but the long-awaited pictures were never published, nor was much else said. Only after *glasnost* did the Russians admit that the mission had failed in its primary purpose and that the pictures had never reached Earth in the first place. Although the cameras had worked, a problem with the thruster systems meant that the spacecraft had not been pointing at the moon at all, but taking pictures of blank space! This was due in turn to a foreign object getting stuck in one of the thrusters, making orientation impossible. Luna 11 also carried instruments to measure gamma rays, X-rays, meteorite streams and hard corpuscular radiation. Specifically, it was instrumented to confirm Luna 10's detection of mascons. The scientific outcomes are not known and few lunar results were attributed to Luna 11. Russian accounts of the scientific results of the 1966 orbiting missions give details of outcomes from Luna 10 and 12, but not 11 [10]. Luna 11 carried, as did its successor, gears and bearings designed to be used on subsequent lunar rovers, to test how they would work in a vacuum.

Luna 12 (22nd October) passed the moon at 1,290 km at a speed of 2,085 m/sec when its retrorocket fired for 28 sec to cut its velocity to 1,148 m/sec to place it into an orbit of 100 by 1,737 km, 3 hr 25 min, in a much narrower equatorial orbit than Luna 11, only 15°. This time, lunar photography was the stated mission objective and

Luna 11, 12 design

presumably this was accomplished on the first day during the low points of the orbital passes. Thrusters were used extensively to point Luna 12 toward landing sites and on the second day the spacecraft was put into a slow roll so as to accomplish the rest of its mission.

The whole mission lasted three months and ended on 19th January 1967 after 85 days, 602 orbits and 302 communications sessions. The imaging, scanner and relay system had a resolution of between 15 m and 20 m and could be transmitted at either 67 lines/frame for 125 sec (quick look) or at 1,100 lines a frame for 34 min (high resolution). The target areas were the Sea of Rains, Ocean of Storms and craters Ariastarcus and Alphonsus: a Soviet photograph released late in 1966 showed cosmonauts Yuri Gagarin, Alexei Leonov, Vladimir Komarov and Yevgeni Khrunov pouring excitedly over its pictures.

The Russians gave only a short account of the Luna 12 mission, the principal one being *Luna 12 transmits*, published in *Pravda* on 6th November 1966 and they released only a small number of images from Luna 12, much inferior in quality to the American

Luna 12 images

lunar orbiters and doing less than justice to the 15 m resolution of the cameras [11]. There are some reports that the photographs were so poor that the Russians ended up resorting to assembling the publicly available American Ranger and Lunar Orbiter archive to plan their moon landings; but this could be a traditional Western under-estimate of Soviet photographic capabilities. There is no suggestion that anything went wrong, so the pictures must have been at least up to the standards of Zond 3. Because they were taken at much closer range, they were probably much better. Either way, it is more than likely that there are still some Luna 12 pictures deep in some Moscow archive. In addition to cameras, Luna 12 carried a gamma ray spectrometer, magnetometer, infrared radiometer and micrometeorite detector. Assessments were made of the reflectivity of the lunar surface to infer its density ($1,400 \, \text{kg/m}^3$).

Presumably, the Luna 12 pictures would have been decisive in determining where the Russians would land on the moon. The American lunar orbiters enabled the Americans to narrow down the choice of the first landing to five prospective sites, all near the equator (likewise, Luna 12 flew over the equatorial belt, between 15°N and 15°S, in a much narrower band than Luna 11, which operated between 27°N and 27°S). A team in the Vernadsky Institute, led by Alexander Bazilsvsky (b. 1937), worked on site selection for the manned landing from 1968 and also for soil sample and rover missions. Eventually, the Russians selected three smooth areas for the first manned landing on the moon:

- Ocean of Storms;
- Sinus Meridiani; and
- Sea of Tranquility (not the Apollo 11 site).

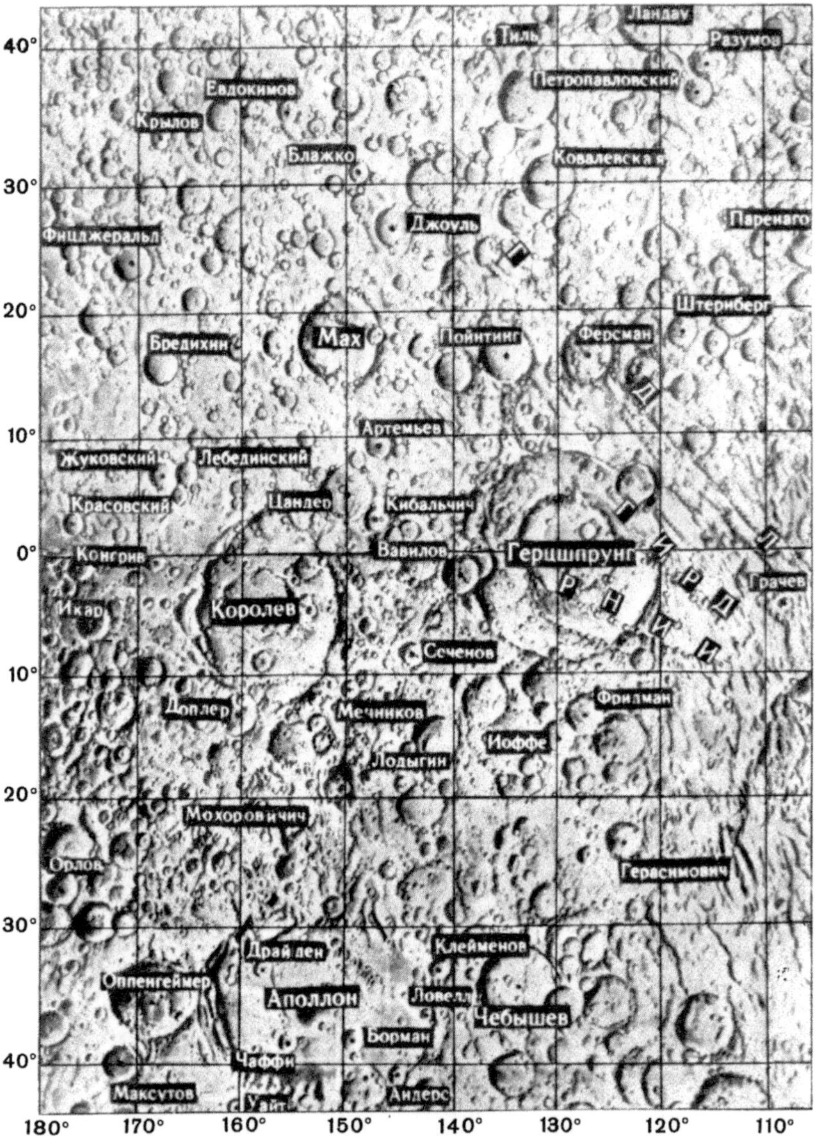

Soviet lunar map

Ye-6LF (originally Ye-7)

Height	2.7 m
Base	1.5 m
Weight (payload)	1,665 kg
Orbiting altitude	100 × 1,700 km
Angle to equator	From 15 to 27°
Orbital period	178 to 205 min

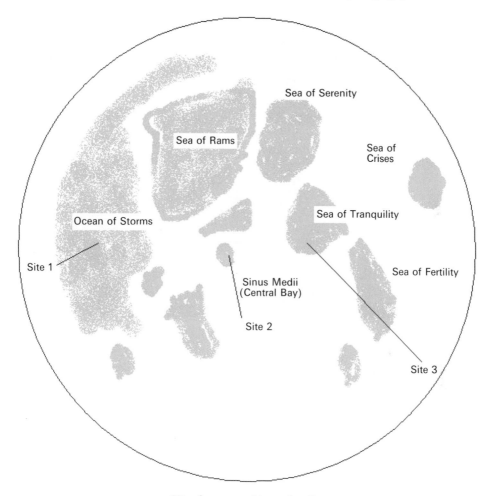

Sites for manned lunar landing

Ye-6 series: instruments specified[1]
Magnetometer.
Gamma ray spectrometer.
Gas discharge counters.
Electrode ion traps.
Meteoroid particle detector.
Infrared radiometer.
Low-energy X-ray photon counter.
Cameras (Ye-6LF).

[1] Cannot be confirmed that all were flown on each mission.

Luna 10 and mother ship

Summary of lunar orbiters
Ye-6S and Ye-6LF

1 Mar 1966	Failure (Cosmos 111)
31 Mar 1966	Luna 10
24 Aug 1966	Luna 11
22 Oct 1966	Luna 12

FINALLY, COMMUNICATIONS TESTS

Russia flew a third series of moon probes. These have no direct American comparator. Called the Ye-6LS, little is known about them. Only an outline sketch of Luna 14 has been released (no photographs), showing that it was similar in design to the Ye-6LF. The purpose of the Ye-6LS series was to test out communications between moon orbit and the deep space tracking network, employing the systems to be used later by the manned lunar orbiter, the LOK. Two scientific instruments were also carried: one to measure charged solar particles, the other cosmic rays. One engineering experiment was also carried: more drive gears and lubricants, to test systems to be used on the upcoming series of lunar rovers.

This programme was sufficiently important for three Ye-6LS probes to be flown. The first was launched on 17th May 1967. It was intended that this spacecraft go into a high-Earth orbit reaching out to the full lunar distance, but away from the direction of the moon. In the event, the fourth stage cut out prematurely, leaving Cosmos 159

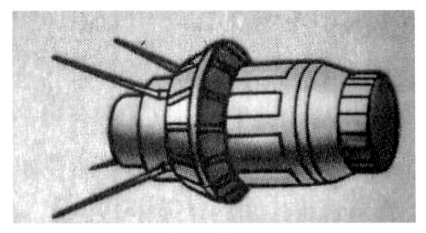

Diagram of Luna 14

in a highly irregular orbit, 260 km by 60,710 km. Although falling far short of that intended, the altitude probably was sufficient for a useful test of the LOK communication systems. The second Ye-6LS mission also failed. Block I cut off prematurely 524 sec into the mission on 7th February 1968 when the fuel inlet control jammed and it fell short of Earth orbit.

The only fully successful mission, Luna 14, was flown out to the moon on 7th April 1968 and entered lunar orbit of 160 km by 870 km, 42°, 2 hr 40 min. Russian news agencies said almost nothing about it, except that it carried out studies of the stability of radio signals and the moon's gravitational field. No one believed them at the time and assumed it was a failed photography mission. In reality, they were telling the truth. It was a test of communications ('radio signals') and a mission designed to measure the perturbation of lunar orbits by mascons. Although the discovery of mascons has always been assumed to be American, in fact it can be attributed to Luna 10. Mascons worried both sides equally, for they pulled orbiting spacecraft out of their predicted orbit by several kilometres. These distortions could make all the difference to where a spacecraft was targeted for landing and to the success of subsequent link-ups in lunar orbit. Apollo 11's *Eagle* nearly ran out of fuel because its targeting was off-course and Neil Armstrong had to fly the lunar module far downrange to find a suitable area. Instruments were also carried to measure solar wind, cosmic rays and charged particles in lunar orbit, although their outcomes were not publicized. The Russians appear to have been well satisfied with Luna 14's radio communications tests and mascon mapping, but, because it revealed much about their manned lunar ambitions, drew little attention to the mission.

Ye-6LS series

17 May 1967	Cosmos 159
7 Feb 1968	Failure
7 Apr 1968	Luna 14

Ye-6S, Ye-6LF, Ye-6LS series: scientific outcomes (with Zond 3)
Very weak magnetic field around the moon (distortion of the interplanetary magnetic field?).
No lunar magnetic poles.
No differences in radiation emission levels between lunar lowlands and highlands.
Cosmic radiation at 5 particles/cm^2/sec.
198 meteoroid impacts (Luna 10).
No gaseous atmosphere around moon found.
Finding of anomalous zones of mass concentrations below lunar surface disturbing the lunar
  orbit (mascons) (Luna 10); characterization of such zones (Luna 14).
Broad composition of lunar rocks (basaltic).
Selection of landing sites for manned and rover landings.
Infrared, ultraviolet scan of lunar surface (Zond 3).
Assessment of reflectivity of lunar surface and inferred density.

ORBITERS, IN CONCLUSION

During 1966–8, the Soviet Union sent up seven orbiters to explore the lunar environment and map the surface. Of these, two failed (Cosmos 111, 7th February 1968) and one partly failed (Luna 11). Only one photographic mission succeeded (Luna 12), but we do not have access to the archive which it assembled. Luna 10, despite being improvised, appears to have returned a substantial amount of scientific information. The Russians also ran a series of communications missions, the Ye-6LS, which shows their thoroughness in approaching the moon project and which have no direct American comparison.

There was discussion, in the Western popular press, as to the need for proceeding to manned flights to the moon when so much useful information had been already retrieved by automatic probes. Later, when the Russians were beaten in the moon race, they raised the question in retrospective justification for their use of automatic probes. In reality, this important discussion was given little airing within the two respective space programmes themselves, for the political decision had already been taken to go for a manned flight around the moon and to its surface. This decision had little to do with a calculation of the best way to obtain a scientific return, but, as President Kennedy himself put it, would be the approach 'most impressive to mankind'. It is back to this larger project that we now turn (Chapter 5).

REFERENCES

[1]   Surkov, Yuri: *Exploration of terrestrial planets from spacecraft – instrumentation, investigation, interpretation*, 2nd edition. Wiley/Praxis, Chichester, UK, 1997.
[2]   Tyulin, Georgi: Memoirs, in John Rhea (ed.): *Roads to space – an oral history of the Soviet space programme*. McGraw-Hill, New York, 1995.
[3]   Ivanovsky, Oleg: Memoirs, in John Rhea (ed.): *Roads to space – an oral history of the Soviet space programme*. McGraw-Hill, New York, 1995.

[4] Clark, Phillip S.: Masses of Soviet Luna spacecraft. *Space Chronicle, Journal of the British Interplanetary Society*, vol. 58, supplement 2, 2005.

[5] Minikin, S.N. and Ulubekov, A.T.: *Earth–space–moon*. Mashinostroeniye Press, Moscow, 1972.

[6] Dohnanyi, J.S.: *Dust cloud produced by Luna 5 impacting on the lunar surface*. NASA, Washington DC, 1965.

[7] Lipsky, Yuri: Major victory for Soviet science – new data on the invisible side of the moon. *Pravda*, 17th August 1965.

[8] Khrushchev, Sergei: The first Earth satellite – a retrospective view from the future, in Roger Launius, John Logsdon and Robert Smith (eds): *Reconsidering Sputnik – forty years since the Soviet satellite*. Harwood Academic Publishers, Amsterdam, 2000.

[9] Leonov, Alexei and Scott, David: *Two sides of the moon – our story of the cold war space race*. Simon & Schuster, London, 2004.

[10] Surkov, Yuri: *Exploration of terrestrial planets from spacecraft – instrumentation, investigation, interpretation*, 2nd edition. Wiley/Praxis, Chichester, UK, 1997.

[11] Luna 12 transmits. *Pravda*, 6th November 1966, as translated by NASA.

Russia's UR-500K Proton

Length	44.34 m
Diameter	4.1 m

First stage (block A)

Length	21 m
Diameter	4.1 m
with tanks	7.4 m
Engines	Six RD-253
Burn time	130 sec
Thrust	894 tonnes
Fuels	UDMH and N_2O_4
Design	OKB-456

Second stage (block B)

Length	14.56 m
Diameter	4.1 m
Engines	Two RD-210, one RD-211
Burn time	300 sec
Thrust	245 tonnes
Fuels	UDMH and N_2O_4
Design	OKB-456

Third stage (block V)

Length	6.52 m
Diameter	4.1 m
Engines	Three RD-213, one RD-214
Burn time	250 sec
Thrust	64 tonnes
Fuels	UDMH and N_2O_4
Design	OKB-456

Fourth stage (block D)

Length	2.1 m
Diameter	4.1 m
Engine	One 11D58M
Thrust	8.7 tonnes
Length	6.3 m
Diameter	3.7m
Fuels	Liquid oxygen and kerosene
Design	OKB-1

THE ROCKET FOR THE LANDING: THE N-1

In contrast to Proton, the N-1's design history has been chronicled in some detail. The N-1 programme began on 14th September 1956, when the first sketches appeared in the archive of OKB-1. Korolev gave it the relatively bland name of N-1, 'N' standing for *Nositel* or carrier, with the industry code of 11A51. The concept was brought to the Council of Chief Designers on 15th July 1957, but it did not win endorsement. The N-1 at this stage was a large rocket able to put 50 tonnes into orbit.

This was in dramatic contrast with the United States, where the Saturn V was constructed around a single mission: a manned moon landing. The N rocket, by contrast, was a universal rocket with broad applications. Korolev kept these purposes deliberately vague and, in order to keep military support for the project, hinting at how the N-1 could launch military reconnaissance satellites. The sending of large payloads to 24 hr orbit was also envisaged.

The N-1 languished for several years. Unlike the R-7, then in development, it did not have any precise military application and as a result the military would not back it. The situation changed on 23rd June 1960 when the N-1 was approved by Resolution #715-296 of the government and party called *On the creation of powerful carrier rockets, satellites, space ships and the mastery of cosmic space 1960–7*. Encouraged by the success of early Soviet space exploration, aware of reports of the developments by the United States of the Saturn launch vehicle, the Soviet government issued a party and government decree which authorized the development of large rocket systems, such as the N-1, able to lift 50 tonnes. The decree also authorized the development of liquid hydrogen, ion, plasma and atomic rockets. The 1960 resolution included approval for an N-2 rocket (industry code 11A52), able to lift 75 tonnes, but it was dependent on the development of these liquid hydrogen, ion, plasma and nuclear engines, suggesting it was a more distant prospect. The 1960 resolution also proposed circumlunar and circumplanetary missions. The implicit objective of the N-1 was to make possible a manned mission to fly to and return from Mars.

Such a mission was mapped out by a group of engineers led by Gleb Yuri Maksimov in Department #9, overseen by Mikhail Tikhonravov. This would not be a landing, but a year-long circumplanetary mission. Maksimov's 1959–61 studies postulated a heavy interplanetary ship, or TMK, like a daisy stem (nuclear power plant at one end, crew quarters at the opposite), which would fly past Mars and return within a year. In 1967–8, three volunteers – G. Manovtsev, Y. Ulybyshev and A. Bozhko – spent a year in a TMK-type cabin testing a closed-loop life support system. A related project, also developed in Department #9 but this time by designer Konstantin Feoktistov, envisaged a landing on Mars using two TMKs. The N-1 design, although it proved problematical for a man-on-the-moon project, was actually perfect for the assembly of a Mars expedition in Earth orbit.

Siddiqi points out that the N-1 Mars project and the development of related technologies took up a considerable amount of time, resources and energy at OKB-1 over 1959–63 [3]. Were it not for the Apollo programme, the Russians might well have by-passed the moon and sent cosmonauts to Mars by the end of the decade. Ambitious though the Apollo project was, Korolev had all along planned to go much farther.

The slogan of the GIRD group, written by Friedrich Tsander, never referred to the moon at all. Instead, its motto was 'Onward to Mars!'

The fact that the original payload was 50 tonnes, far too little for a manned moon mission, demonstrates how little a part a moon landing played in Korolev's plans at this time. Indeed, unaware of its origins, several people later questioned the suitability of the N-1 for the moon mission. Sergei Khrushchev said it was neither fish nor fowl, too large for a space station module, too small for a lunar expedition [4]. The head of the cosmonaut squad, General Kamanin, took the view in the mid-1960s that the design, going back to 1957, was already dated. The rival and ambitious Vladimir Chelomei felt he could do better with a more modern design.

Despite approval in 1960, the N-1 made very slow progress, and early the following year Korolev was already complaining that he was not getting the resources he needed. At one stage, in 1962, the government halted progress, limiting the N-1 project only to plans. An early problem, one that was to engulf the project, was the choice of engine. Korolev needed a much more powerful engine for such a large rocket. Korolev proposed kerosene-based engines for the lower stages and hydrogen-powered engines for the upper ones. He turned, as might be expected, to Valentin Glushko of OKB-456, asking him to design and build such engines. Valentin Glushko proposed a series of engines for the different stages of the N-1: the RD-114, RD-115, RD-200, RD-221, RD-222, RD-223. None of these was acceptable to Korolev, for all were nitric-based, anathema to him [5].

The ever-resourceful Korolev then turned to Kyubyshev plane-maker Nikolai Kuznetsov, asking him to make the engines using the traditional fuels of kerosene and liquid oxygen. Although Kuznetsov had no experience of rocket engines, he and his OKB-276 design bureau were prepared to give it a try. However, he knew he had no ability to develop high-powered engines, so a large number of modest-power kerosene-fuelled engines would have to do. In fact, despite his inexperience, the engines came out exceptionally powerful and lightweight, achieving the best thrust-to-weight ratios of the period. The engines were called NK, NK standing for Nikolai Kuznetsov. The first stage used NK-33s, the second NK-43s. There was little difference between them except that NK-43 was designed for higher altitudes and had a larger nozzle [6]. Roll was controlled by a series of roll engines. For example, on the first stage, there were four roll engines of 7 tonne thrust each, assisted by four aerodynamic stabilizers.

First-stage separation would take place at 118 sec, at 41.7-km altitude, by which time the rocket would be travelling at 2,317 m/sec. The second stage would burn for over a further two minutes, reaching 110.6 km, with speed now at 4,970 m/sec. The third stage would then burn until 583 sec, by which time the stack had reached orbital altitude of 300 km and a velocity of 7,790 m/sec. Translunar orbit injection would be done by block G. This would use a single NK-31 engine, burning for 480 sec, to fire the stack moonward. Block D would then be used three times:

- Lunar orbit insertion (110 km).
- Lunar orbit adjustment (110 km by 16 km).
- Descent to the moon.

Mstislav Keldysh

The N-1 was originally designed to have hydrogen-powered upper stages. Research on hydrogen engines dated to 1959 in Arkhip Lyulka's OKB-165 and 1960 in Alexei Isayev's OKB-2. The main engine design bureau was that of Valentin Glushko, but he had no time for hydrogen-fuelled stages. In May 1961, Korolev contracted Lyulka to build a 25 tonne thrust engine and Isayev a small one of 7 tonne thrust. Within two years, they were able to come back to him with the specifications of their motors, called respectively the 11D54 and 11D56. However, their progress was slow. A critical factor was the lack of testing facilities. Although both tried to get the use of the main testing facility at Zagorsk, priority was given to the testing of military rockets. In the event, they were not able to conduct the necessary tests until 1966–7. By 1964, Korolev had abandoned hope of getting hydrogen-powered engines available on time for the N-1 and went for more conventional solutions.

By summer 1962, the N-1 approached a critical design review. The N-1 design was studied by a commission presided over by Mstislav Keldysh for two weeks in July 1962. The Keldysh Commission gave the go-ahead for the N-1, with Korolev's choice of liquid oxygen and kerosene engines. Siddiqi [7] points to the significance of this decision, for it forever fractured the Soviet space programme into rival camps: Korolev's OKB-1 on one side; and on the other, Glushko's OKB-456 and Chelomei's OKB-52. The payload was set at 75 tonnes, merging the N-1 and N-2 design concepts (another interpretation is that the N-2 was in effect renamed and superseded the N-1). Either way, 75 tonnes was now the base line.

Two months later, on 24th September 1962, a government decree called for a first test flight in 1965. The N-1 was once again given the green light. Despite these changes, the N-1 and its engines continued to make very slow progress. Promised funding never arrived and despite seven years of design and redesign, no hardware had yet been cut. When the Soviet Union began to respond to the challenge of Apollo, Korolev saw the moon landing as an opportunity to give the N-1 the prominence he believed it deserved. On 27th July 1963, Korolev wrote a memo confirming that the N-1 would now be directed toward a manned landing on the moon. Now that the nature of American lunar ambitions had became more apparent, the N-1 was directed away from Mars and toward the moon. His first ideas for a lunar mission for the N-1 were to use two N-1s for his manned lunar expedition, employing the technique of Earth orbit rendezvous. The precise point at which Korolev moved from the Earth orbit rendezvous profile to the lunar orbit rendezvous profile is unclear. Granted the difficulties they had experienced with getting funding for the N-1, the prospect of having to build only one, rather than two, for each moon mission must have been appealing. The favourable trajectory and payload economics of the American lunar orbit rendezvous method persuaded him that a single N-1 could do the job, but it would have to be upgraded again, this time from a payload of 75 tonnes to one of 92 to 95 tonnes, almost double its original intended payload.

When the decision to go to the moon was taken on 3rd August 1964, the N-1 was designated as the rocket for the lunar landing programme. The 95 tonne requirement had two immediate implications. First, the number of engines must be increased from 24 to 30, giving it a much wider base than had originally been intended. The 24 were in a ring and the additional six were added in the middle. Second, the fuel and oxidizer tanks would necessarily be very large. Korolev decided that spherical fuel tanks were to be used, eschewing the strapping of large fuel tanks to the side of the rocket. The diameters in each stage would be of different dimensions, making the system more complex and meaning that the rocket would be carrying a certain amount of empty space. The largest tank was no less than 12.8 m across! Korolev's fuel tanks were so huge that they could not be transported by rail and had to be built on site at the cosmodrome. The first stage had 1,683 tonnes of propellant, of which no fewer than 20 tonnes were consumed *before* take-off!

Trying to get a 95-tonne payload out of a 75-tonne payload rocket design was quite a challenge. Other economies were sought and changes made:

- Setting a parking orbit of 220 km, lower than the 300 km originally planned.
- Additional cooling of fuels prior to launch.
- Thrust improvements of 2% in each engine.
- Use of plastic in place of steel in key components.
- Parking orbit inclination from 65° to 51.6° (later 50.7°).
- Reducing the crew of the lunar expedition from three to two.

Many different – and sometimes rival – branches and bureaux of the Soviet space industry were involved in the N-1 and the programme to put a Soviet cosmonaut on the moon. The N-1 was a huge industrial scientific undertaking, employing thousands of people in Kyubyshev, Moscow, Dnepropetrovsk, Baikonour and many other locations. These were some of the main ones:

Builders of the N-1

Bureau	Chief designer	Responsibility
OKB-1	Korolev, then Mishin	Overall management, block G and D, engines for D, LOK
OKB-276	Nikolai Kuznetsov	Engines for blocks A, B, V
OKB-586	Yangel	LK (spacecraft and engine)
OKB-2	Isayev	LOK propulsion systems
NII-94	Viktor Kuznetsov	Guidance systems for block D, LOK and LK
NII-AP	Pilyugin	Guidance system for LOK
NII-885	Ryazansky	Radio-telemetry systems
GSKB	Barmin	Launch complex
OKB-176	Archip Lyulka	Engines for blocks G, V

The overall designer was Sergei Kryukov, a graduate of the Moscow Higher Technical School, one of the experts sent to Germany in 1945 and a collaborator with Korolev on the R-7.

BUILDING THE N-1

The revised 95-tonne N-1 design was frozen and signed off by Korolev on 25th December 1964. There was far from unanimity on the design and several knowledge-able engineers insisted that the original 75 tonne N-1 design had been pushed beyond its natural limits, with a consequent risk of failure. The piping was a plumber's nightmare. The launch mass was an enormous 2,700 tonnes. Although the N-1 was to be the same size as America's Saturn V, its less efficient fuels produced a smaller payload. Even though the N-1 would follow a profile identical to Saturn V and Apollo, N-1 had the capacity to send only two cosmonauts into lunar orbit and only one down to the surface.

Although the project had first been mooted in 1956 and approved in 1960, sig-nificant resources did not begin to flow into the project until late 1964. Now the N-1 had to catch up for time lost. In his effort to do so, Korolev took two important decisions:

- Although the rocket would be built in Kyubyshev and Moscow, it would be assembled and integrated at Baikonour Cosmodrome, saving transit time.
- Savings would be made on ground testing. Although engines would be tested individually, there would be no testing of all the first-stage engines together on a dedicated test stand. This was in dramatic contrast with the United States, where the large new F-1 engines were tested in large-scale facilities in Huntsville, AL.

Spurring people on – Korolev in 1964

Korolev's philosophy was to fly rockets at the first available opportunity, so that whatever flaws might be there revealed themselves early on during a rocket's development (most rockets were happy to oblige). Under the intense pressure to get the N-1 programme under way, Sergei Korolev and his successor Vasili Mishin minimized the ground-testing segment, a mistake the Americans did not make with the Saturn V (nor the Chinese in their programme). Korolev sometimes said in his defence that the government would not pay for proper ground-testing facilities and they would have held him up too much. Khrushchev does not agree and says that the government would never have denied Korolev proper ground-testing facilities had he asked for them – 'especially a chief designer of Korolev's calibre' [8]. Chelomei's Proton, though, presents a counter-argument. It was extensively ground-tested, but its flight record until 1972 was no better than some of Korolev's rockets.

The N-1 engine system

In an effort to compensate for possible first-stage problems, a special control system was introduced by Vasili Mishin. There was a real danger that the failure of an individual engine could jeopardize the whole mission. Accordingly, Mishin designed an engine operation control system, called KORD (*Kontrol Roboti Dvigvateli*) in Russian, which would shut down any badly performing engine automatically and a good engine immediately opposite, so as to preserve the symmetry of the vehicle's thrust. KORD would also re-programme the burn so as to make up for the lost thrust. The system could tolerate the loss of four first-stage and two second-stage engines and still achieve orbit.

Compared with the Gas Dynamics Laboratory, Kuznetsov's OKB-276 was poorly resourced and lacked *any* rocket engine testing facilities when it was awarded the contract for the moon rocket. Korolev was adamant about all-up testing and would never launch dummy upper stages, which he regarded as wasteful (the

Americans took a different view). Korolev argued that there was a high degree of commonality between all the NK engines and if they were tested properly individually, they should work in stages. The KORD system should be able to cope with any problems arising and the redundancy built in should protect against catastrophic failure. Against that, all engine designers will argue that exhaustive ground-testing will reveal old flaws, new flaws and be worth the investment. The preparation of the N-1 was such a huge event that the programme could not survive repeated catastrophic failures (indeed, it didn't).

There were no ground tests of all the first-stage engines together, with or without the KORD system. But it is wrong to say that there were no ground tests at all. To the contrary, the main rocket-testing centre in the Soviet Union, at Zagorsk (now Sergeev Posad) was used for otherwise exhaustive engine tests over 1967–8. The first engine tests began in September 1967. The NK engines were erected on stands and tested for long periods. The second, third, fourth and fifth stages were tested as stages there, but not, of course, the full first stage. Granted the commonality between the engines, they may reasonably have hoped that mastering the first stage should not present serious difficulties. By the time of the first flight of the N-1, the engines had been tested for 100,000 sec on the ground.

At last, with the government resolution of August 1964, there was a sudden surge of momentum. Construction of the first N-1 pad began the next month in September. After blueprints were agreed on 30th December 1964, the government issued a decree for the construction of 16 N-1 rockets on 26th January 1965. Construction of a second, adjacent pad began in February 1966. The first N-1 hardware appeared in Baikonour in November 1966. The first pad was completed in August 1967. A mockup of the N-1 was rolled out there on 25th November 1967. This was a wise precaution, for there were many problems in trying to integrate the model with the launch pad and its fuelling systems. These took a year to sort out. Two weeks later, on 11th December, it was spotted by an American Corona photoreconnaissance satellite. The model threw a shadow across a quarter of a mile that late afternoon and set alarm bells ringing in Washington DC [9].

These pads were the largest project ever made at Baikonour. Concrete pads were built, with flame trenches gouged out of the ground underneath. Two pads were built, 500 m apart. Each had a 145 m tall rotating servicing tower and 42 m deep flame trenches. Propellants were supplied to the towers by a fuel area in between the two. To get the N-1 rocket to the pad, two railway lines were constructed from the hangar. Two diesel engines moved the empty N-1 down to the pad in parallel. Once it reached the pad, a giant crane erector moved the rocket to the vertical. The N-1 was assembled in what must have been the largest structure built at Baikonour to that point, the final assembly building. This was a huge hangar 240 m long and 190 m wide, its height ranging from 30 m to 60 m, with two high bays and three low bays [10].

The promised Lunar Exploration Council or LEK now set about its business and the N-1 passed a further design review under Keldysh on 16th November 1966. The government resolution on the moon programme of 4th February 1967 laid down a schedule for the first twelve flights of the N-1. The schedule was ambitious, to the point

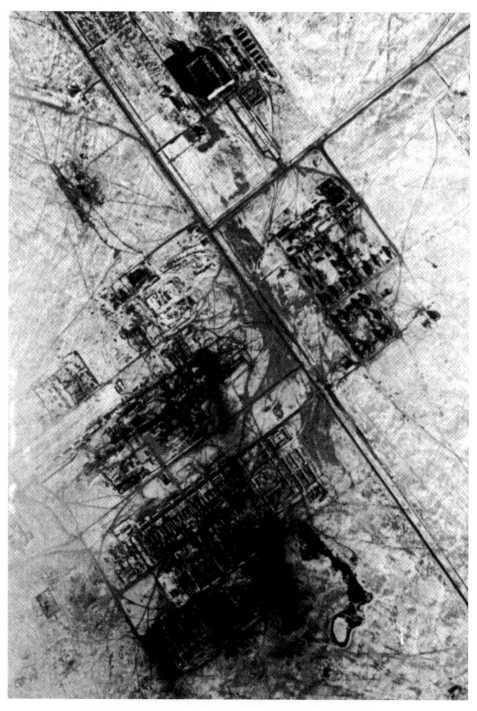

N-1 pads under construction, as seen by American intelligence

of being wholly unrealistic. The plan specified an unmanned test in September 1967 (about the same time as the American Saturn V), leading to a manned lunar landing in September 1968.

The first complete N-1 rocket began construction in Samara in February 1967. Two flight models reached an advanced stage by September 1967. The first N-1 was rolled out to the pad on 7th May 1968. By then, the first N-1 launch was scheduled for later that year. Cracks were then spotted in the first-stage tanks and the whole rocket was removed over 10th–12th June. This was quite a setback to the schedule and meant that the N-1 would not fly in 1968. To continue the momentum of the programme, a mockup N-1 was at the pad from August to October for launchpad tests. To the snooping American satellites, this provided a prolonged opportunity to photograph the target during a period of good sunlight and enable the Americans to model the launcher. But even with their advanced photography, from their altitude of 150 km, the difference between a mockup and the real thing may not have been obvious.

N-1

Length	105 m
Diameter	17 m
Liftoff weight dry	281 tonnes
Liftoff weight fuelled	2,750 tonnes
Payload	95 tonnes

First stage (block A)

Thrust	154 tonnes each
Engines	30 NK-33
Thrust (each)	154 tonnes
Fuels	Liquid oxygen, kerosene
Total thrust	4,620 tonnes
Height	28 m
Diameter	17 m
Weight	1,875 tonnes
Burn time	118 sec

Second stage (block B)

Height	20 m
Diameter	7.5 m
Engines	Eight NK-43
Thrust (each)	179 tonnes each
Total thrust	1,432 tonnes
Fuels	Liquid oxygen and kerosene
Burn time	130 sec
Weight	540 tonnes

Third stage (block V)

Height	12 m
Diameter	6 m

Engines	Four NK-39
Thrust (each)	41 tonnes
Total thrust	164 tonnes
Weight	185 tonnes
Fuels	Liquid oxygen and kerosene
Burn time	400 sec

Fourth stage (block G)

Engines	Four NK-31
Fuel	Liquid oxygen and kerosene
Height	8 m
Diameter	6 m
Total thrust	41 tonnes

Fifth stage (block D)

Engine	One Melinkov RD-58
Fuel	Liquid oxygen and kerosene

Few people have made direct comparisons between the Saturn V and the N-1. One who has is Berry Sanders [11]. His findings were that:

- The Saturn V had a steeper trajectory than the N-1, which rolled on its side sooner.
- The Saturn V was bigger and more powerful. The first stage of the Saturn V had a fuelled weight of 2,244 tonnes, compared with the 1,875 of the N-1. On the Saturn V, the main effort at lifting was done by the hydrogen-powered second stage, while on N-1 the burden was spread evenly between the three stages. Hydrogen gave the Americans a definite advantage.
- Saturn V could put 117 tonnes into orbit, compared with 95 tonnes for the N-1. However, the N-1 paid a considerable penalty for launching into a 51° orbit from a launch site as far north as Baikonour. Had the N-1 been launched from Cape Canaveral, the N-1 could have put 104 tonnes into orbit.

Evaluating the N-1 is a difficult undertaking. At first sight, it was a disastrous rocket, for it exploded four times out of four. By comparison, the Saturn V was an engineer's dream, for it flew 13 times and succeeded 13 times (not that all launches were incident-free, but they all made it into orbit). After the programme was over, Kuznetsov continued work on his NK-33 engine at his own expense. He decided on a duration test of 20,360 sec on a test stand. It ran perfectly. Fourteen engines logged up to 14,000 sec in other tests. Chief designer Mishin considered it the best rocket engine ever made.

Kuznetsov's engineers received the orders from Valentin Glushko to destroy the engines in 1974, but they could not bring themselves to do it. Instead, they hid them in a shed and put a big nuclear skull-and-crossbones warning sign over them, believing that would keep prying eyes away, which it did. There the engines gathered dust for 20 years and were rediscovered, almost by accident. Visiting American engineers saw them on a visit to Samara and could not believe their eyes: hundreds of moon rocket engines in mint order! The American Aerojet company at once bought 90 of them for $450 million and in 1995 sent them off to its Sacramento, CA plant for testing and evaluation. They worried if there would be any problems in relighting motors that had been in storage since 1974. They ran two tests – of 40 sec and 200 sec – and there were not. Aerojet's evaluation of the engine found that it could deliver over 10% more performance than any other American engine and enthused over its simplicity, lightness and low production costs. The hydrogen upper stage, originally planned for a later version of the N-1, became the upper stage of the Indian GSLV launch vehicle more than 25 years later.

The basic problem with the N-1 was the lack of thorough ground-testing. It was here that the much smaller resources of the Soviet Union and poor organization told against its moon programme. Rocket designers continued to underestimate the problems associated with the integration of engines on stages and the resulting problems of vibration, sound, fuel flow and control. Testing engines individually, however good they are, can be a poor guide as to how they behave collectively. Even where this is done, there is no guarantee of success, as the thoroughly prepared Proton proved. The world's space programmes are full of histories of rockets that proved extraordinarily difficult to tame: the American Atlas and Centaur, the Chinese Feng Bao, Europe's Europa and India's SLV. Even programmes that have built on the experience of all that has gone before can suffer nasty surprises, like Europe's Ariane. Having said all that, it is hard to believe that thorough ground-testing would not have stacked the odds much more in favour of the N-1. The four flight failures all had their roots in problems that could have been identified in thorough ground-testing. The real issue is not that the N-1 was a bad rocket, but that the Saturn V was so exceptionally good.

Were the rival UR-700 and R-56 proposals better? The UR-700 scheme developed by Chelomei might well have worked. In promising exhaustive ground-testing first, Chelomei rightly hit on one of the great weaknesses of Korolev's approach. Chelomei was a superb designer but he was also slow: his Almaz space station was approved in 1964 but he did not get it ready for its first flight until 1973 and there is no reason to believe he could have built his moon rocket any sooner.

In retrospect, the Russian moon programme might have been better to go for the large RD-270 engines which Glushko began to develop. Korolev probably correctly judged that the development of the RD-270s would have required an extensive range of ground facilities and taken too long. With time against him, he calculated that it was better to go with a tried-and-tested system, even if it meant 30 engines. Korolev was always battling time to get his N-1 airborne, struggling with government departments for budgets and travelling endlessly to Samara, Leningrad and Baikonour to keep things moving. Korolev may have reckoned that he had to be lucky with only one successful N-1 launch and he would then get, from the political bosses, the resources he needed to bring the project to fruition.

If the N-1 had eventually worked, then a Russian moon landing would definitely have been possible at some stage. Alternatively, if the Russians had decided not to pursue a moon landing, then they would have had available to them a large rocket able to launch a very big space station. Several such designs were even sketched during this period. Instead of the smaller, Mir-class space station of 1986–2001, the Russians would have been able to put in orbit a large space station block long before 1980, something as large as the International Space Station. The cancellation of the N-1 not only marked the effective end of the Soviet man-on-the-moon programme but had a profound effect on the subsequent development of cosmonautics.

A SPACESHIP TO CIRCLE THE MOON

The original around-the-moon programme was designed by Tikhonravov's Department #9 of OKB-1 in 1960–1, and this became the Soyuz complex of 1962–4. In August 1964, the around-the-moon programme was transferred to Vladimir Chelomei's OKB-52 design bureau. He planned to send a spacecraft, called the LK (*Luna Korabl*) directly around the moon on his Proton rocket, then nearing completion. The idea that the Soviet Union might attempt to send a man around the moon first was one familiar to Western analysts. The around-the-moon mission required much less rocket power, hardware and testing than a landing. The psychological effect of going around the moon, the excited commentaries, in Russian, of the lunar surface at first hand, would have a considerable effect on world public opinion. Chelomei probably realized this.

Not much is known of Chelomei's LK design. A design published in the Tsiolkovsky Museum in Kaluga shows a bullet-shaped cabin with two solar wings at the base, eight aerials and a service module of some kind behind. It resembled a scaled-down Apollo-type command-and-service module, 5.2 m long with 7.27 m wide solar panels and X-shaped antenna system, possibly 4 tonnes in weight. The small, 2.7 m long 2 tonne cabin would have carried one person around the moon. Fitted to the top of the UR-500K, the entire space vehicle would have been 46.7 m tall. The design, completed in July 1965, seems to have made little progress, and it is possible that Chelomei, like Korolev, was severely overstressed with other projects, in Chelomei's case the development of the Almaz orbital space station. Vladimir Chelomei was an original and imaginative designer who came up with many ingenious designs and solutions and it is possible that his LK might have been one of them. Even today, many years after his death in 1984, his influence is still apparent. His design, the Proton, is still flying, a new version being introduced, the Proton M. The first module in the International Space Station, the functional control block or *Zarya*, is originally a Chelomei design.

Chelomei's LK design was to become an academic matter. In October 1964, only a few months after the August governmental resolution, Nikita Khrushchev was overthrown. Khrushchev had been a big supporter, largely because of Chelomei's success in delivering a fleet of operational ballistic missiles for the Soviet rocket forces.

Korolev devoted considerable energies during 1965 trying to push Chelomei out of the moon programme altogether and instead for OKB-1 to run an integrated programme for around-the-moon voyages and landing, which he argued made more economic and organizational sense. Eventually, on 25th October 1965, Korolev managed to wrest the LK moonship back from the Chelomei design bureau. Korolev was able to offer a stripped-down Soyuz spacecraft as his alternative, which he called the 7K-L-1. The government must have been persuaded that a design that was already at an advanced stage was preferable to one that had barely got beyond the drawing board. Korolev was not able to remove Chelomei altogether, for the government decided that the UR-500 would continue to be used. Korolev also persuaded the government to use, as upper stage for the Proton, the block D upper stage then being fitted out for the N-1 rocket. On 31st December, Korolev and Chelomei formally signed off on the deal.

It would be wrong to overstate the rivalry between Chelomei and Korolev, for they seemed able to work together when it mattered, albeit sullenly on Chelomei's part. This was not the case between Korolev and Glushko, whose relationship seems to have become truly venomous. With the man-around-the-moon project using the same block D upper stage and a related cabin, the 7K-L-1, the Soviet moon programme was at last achieving some economies of scale. The December 1965 agreement specified the construction of no fewer than fourteen L-1 spacecraft, of which seven would be for unmanned tests and four for manned circumlunar missions.

Both the Russian moonships, the L-1 Zond and the LOK, were derivatives or relatives of the Soyuz spacecraft, which in turn was rooted in the designs of the Soyuz complex, 1962–4. The missions of the L-1 Zond and LOK were closely, even intimately, linked to the development of Soyuz.

ZOND'S ANCESTOR: SOYUZ

The basic Soyuz was 7.13 m long, 2.72 m wide, with a habitable volume of $10.5\,m^3$, a launch weight of up to 6,800 kg, and a descent capsule weight of 2,800 kg. Soyuz consisted of three modules: equipment, descent and orbital. The equipment module contained retrorockets and manoeuvring engines, fuel, solar wings and supplies. The acorn-shaped descent module was the home of cosmonauts during ascent and descent, which one entered through the top. There were portholes, a parachute section and three contour seats. The orbital module, attached on the front, was almost circular, with a spacewalk hatch, lockers for food, equipment and experiments. Being more spacious, the cosmonauts lived there rather than the cramped descent module. From Soyuz there protruded a periscope for dockings, two seagull-like solar panels, aerials, docking probe on the front and flashing lights and beacons. On top of the Soyuz was an escape tower. Normally jettisoned at 2 min 40 sec into the flight, the purpose of the escape tower was to fire the Soyuz free of a rogue rocket. A solid rocket motor, with twelve angled nozzles of 80,000 kg thrust, would fire for 5 sec.

The Soyuz spacecraft

The initial tests of Soyuz were not auspicious. The first test of Soyuz was Cosmos 133 on 28th November 1966. Cosmos 133 was to have docked with a second Soyuz, launched a day later, but this launch was cancelled when Cosmos 133 developed attitude control problems. The Cosmos could not be positioned properly for reentry and was destroyed deliberately for fear that it would land in China. During the second test, a month later, the rocket failed to take off. When the gantries were swung back around the rocket, the cabin was accidentally tipped, causing the escape tower to fire, thus setting the upper stage on fire and causing an explosion which destroyed the pad. One person died, but it could have been many more. The third test, Cosmos 140 on 7th February 1967, followed the test profile up to reentry when a maintenance plug in the heatshield burned through and caused structural damage. Worse followed: the cabin came down in the Aral Sea, crashed through ice and sank (divers later retrieved the cabin from 10 m down).

Despite these difficulties, Russia pressed ahead with a first manned flight of the Soyuz for April 1967. Instead of a cautious, single mission, a big shot was planned. Soyuz would go first, with a single cosmonaut on board, Vladimir Komarov. Twenty-four hours into the flight, Soyuz 2 would follow, commanded by veteran, Valeri Bykovsky. Two newcomers, Yevgeni Khrunov and Alexei Yeliseyev would fly with him. The rendezvous would simulate the moon link-up. Soyuz would be the active craft and would rendezvous on orbit 1. Then the show would really begin. Khrunov and Yeliseyev would don suits, leave Soyuz 2 and transfer into Soyuz to join Komarov. The spacewalk would simulate the transfer of cosmonauts between the the lunar orbiter and lunar lander as they circled the moon. The two ships would then separate after about four hours. Komarov, now accompanied by Khrunov and Yeliseyev, would be back on the ground by the end of day 2, Bykovsky following on day 3. So, in 72 breathtaking hours, the new Soyuz craft would demonstrate Earth orbit rendezvous on the first orbit, transfer by spacewalking to a primitive space station, carry out key tests for the moon flight and put the USSR back in front.

As the launch date drew near, there were a record 203 faults in Soyuz which required correction. The pre-test flights had been disconcerting. An atmosphere of foreboding prevailed at the cosmodrome. As Vladimir Komarov climbed into the transfer van to take the ride down to the pad, he had an air of fatalistic resignation about him. His fellow cosmonauts joshed him, trying to cheer him and get a smile. They started singing, encouraging him to join in. By the time they reached the pad some minutes later, he was singing with them too and the mood of pessimism had lifted somewhat. At 3:35 a.m. Moscow time (not quite sunrise local time) on 23rd April 1967, the R-7 rocket lit the sky up and headed off in the direction of the growing embers of the onrushing dawn. Eight minutes later Vladimir Komarov was back in orbit testing out the most sophisticated spacecraft ever launched.

The trouble started at once when one of Soyuz's two solar panels failed to deploy, starving the craft of electrical power. Other glitches developed as the day went on. The first attempt to change the craft's orbit was unsatisfactory. The ship began to rotate around its axis and only spun more when Komarov tried to correct the problem. The thermal control system degenerated, communications with the ground became irregular and lack of electricity prevented the astro-orientation system from operating. The ion system had to be used instead. Ground control was considering a way of launching Soyuz 2 and for the spacewalking cosmonauts to free the errant solar panel of Soyuz when a tremendous storm hit the launch site and knocked out the electrical systems of the waiting rocket. The decision was taken to abandon the Soyuz 2 launch and bring Komarov home at the first available opportunity, on orbit 16 the next morning.

Even then, there was more trouble. Just as the attitude control system was lining up the Soyuz for reentry, the craft passed into darkness and it lost orientation. The decision was made to try again on orbit 17, even though it too would bring Soyuz far away from the normal landing site. Using procedures that he had never practised in training, Komarov managed to align the craft and fire the retrorockets himself. Despite his heroic efforts to save the mission, worse was to come. As the cabin descended through the atmosphere, the drogue parachute came out but the main parachute remained stubbornly in its container. When the reserve chute was popped out, it tangled in the lines of the drag chute of the main parachute. Soyuz 1 crashed at great speed into the steppe at Orenberg at 7 a.m. The cabin exploded on impact and when Air Force recovery teams arrived all they found was burning metal, the rim of the top of Soyuz being the only hardware they could identify. They piled on soil to extinguish the flames.

The control centre knew nothing of what had happened. As they closed in on the wreckage, the recovery team sent a garbled message to the effect that the cosmonaut needed 'urgent medical attention' (a euphemism for the worst possible news), but the local Air Force commander closed off all communications. Defence Minister Ustinov was informed of the true outcome at 11 a.m. and Leonid Brezhnev an hour later in Karlovy Vary, Czechoslovakia. The Soviet people were officially informed later in the day. Gagarin himself removed Komarov's body from the wreckage. Some days later, some young Pioneers (boy scouts) found some further remains of Vladimir Komarov on the steppe. They buried them and made a small memorial for him of their own.

Soyuz spacecraft rendezvous, docking system

Vladimir Komarov's loyal comrades laid his remains to rest in the Kremlin Wall two days later. It was a sombre and chilling occasion, an unwelcome reminder of the real costs of the moon race. As the bands played the haunting Chopin funeral march the grim-faced and tight-lipped cosmonaut corps, now diminished to nine men and one woman, swore that the programme must go on relentlessly.

The consensus afterwards was that the whole mission had been rushed before Soyuz was really ready. It was apparent that Komarov had behaved masterfully in steering Soyuz successfully through reentry against all the odds. The failure in the parachute system was quite unrelated to the many problems that had arisen in the flight up to that time. The system for sealing the parachute container was defective, making the parachute likely to stick as it came out. This left the investigators with the chilling conclusion that if Soyuz 2 had been successfully launched, it too would have

Vladimir Komarov and his friend Yuri Gagarin

crashed on its return. Years later, Valeri Bykovsky recalled how the storm had saved his life.

Early tests of Soyuz

28 Nov 1966	Cosmos 133 (failure)
Dec. 1966	Pad explosion
7 Feb 1967	Cosmos 140 (failure)
23 Apr 1967	Soyuz (failure)

THE SPACESHIP TO CIRCLE THE MOON: THE L-1, ZOND

Preparations for the flight of Soyuz coincided with those of the Soyuz-derived L-1 cabin, which would fly a cosmonaut around the moon. The L-1 cabin was later called Zond, thus creating confusion with the engineering tests developed by Korolev as part of the interplanetary programme. Zond 1 had flown to Venus, Zond 2 to Mars, Zond 3 to the moon to test equipment for Mars and now Zond 4–8 would fill an important part in preparations to send cosmonauts around the moon.

From August 1964, the Soviet lunar programme had been divided between the around-the-moon programme (Proton, L-1/Zond) and the manned lunar landing (N-1, LOK, LK). When the State Commission on the L-1 met in December 1966, it set a date for the first manned circumlunar flight of 26th June 1967, to be preceded by four unmanned tests.

Zond was a stripped-down version of Soyuz. Its weight was 5,400 kg, length 5 m, span across its two 2 m by 3 m solar arrays 9 m, diameter 2.72 m and a habitable volume of 3.5 m^3. It could take a crew of either one or two cosmonauts in its descent module. The sole engine was the 417 kg thrust Soyuz KDU-35 able to burn for about 270 sec, but it fired more thrusters than Soyuz. Its heatshield was thicker than Soyuz in order to withstand the high friction on lunar reentry at 11 km/sec. It carried an umbrella-like, long-distance, high-gain antenna and on the top a support cone, to which the escape tower was attached. Designer was Yuri Semeonov. The following were the main differences between Soyuz and L-1 Zond. Zond was:

- Smaller, without an orbital module.
- Maximum crew of two, not three.
- Instrument panel configured for lunar missions.
- Support cone at top.
- Long-distance dish aerial for communications.
- Thicker, heavier heatshield for high-speed reentry.
- Removal of docking periscope.
- Smaller solar panels.

Theoretically, Zond could take three cosmonauts; but, without an orbital module it would be tight enough for two for the 6-day mission. The mission profile was for Proton to launch Zond into a parking orbit. On the first northbound equator pass, block D would ignite and send Zond to the moon. Zond would take three days to reach the moon, swing around the farside in a figure-of-eight trajectory and take three days to come home. The spacecraft would have to hit a very narrow reentry corridor. It would use Tikhonravov's skip technique to bounce out of the atmosphere, killing the speed and then descend to recovery. The designers decreed that there should be four successful missions out to the moon, or a simulated moon, before putting cosmonauts on board for the mission.

An important distinction between Apollo on the one hand and the Soyuz, L-1/Zond and LOK on the other was the high level of automation on Soviet spacecraft. As noted above, the Soviet Union decided to pave the way for a manned flight around the moon with no fewer than four automatic flights that precisely flew the same profile as the intended manned spacecraft. A comparable régime would have been followed for the LOK and a series of wholly automatic tests were set for the lunar lander, the LK and the block D upper stage. The Soyuz system was designed to achieve entirely automated rendezvous and docking in Earth orbit. All this required a high level of sophistication in control and computerized systems, something the Russians were rarely given the credit for. From the start, Korolev had built a high degree of automation into spacecraft, a decision which seems to have gone unchallenged

Cosmonaut Valeri Kubasov in Zond simulator

and the early manned spaceship designs were finalized long before the first cosmonauts arrived. The head of cosmonaut training, General Kamanin, is known to have been privately critical of the high level of automation and the lack of scope given to cosmonauts to fly their own spacecraft. Zond carried the first computers used on Soviet spacecraft, the *Argon* series. *Argon* weighed 34 kg, light for its day and was the primary navigation system. It was assessed as having a reliability rate of 99.9% [12]. The Argon 11S was completed in 1968 in time for the Zond lunar missions. The cosmonauts would control the L-1 with a command system called *Alfa*, which had a 64-word read–write menu and 64 commands, with a choice of 4,096 words [13].

Although Zond was based on Soyuz, it had an entirely different control panel. As happened from time to time in the Soviet space programme, this came to light by accident. During the late 1970s, at a time when the Russians claimed 'there had never been a moon race', they released pictures of cosmonauts Vladimir Shatalaov and Valeri Kubasov in training, set against a background of what was presumed to be a Soyuz control panel. It must have escaped the censors that the control panel was entirely different from the Soyuz, with the Earth orbit orientation system taken out. The Zond control cabin comprised a series of caution and warning panels; cabin pressure, composition and electric meters; computer command systems; periscope; and translunar navigation systems.

The Soviet around-the-moon mission operated under a number of constraints, as follows:

- The moon should be high in the sky over the northern hemisphere during the outward and returning journey, so as to facilitate communications between Zond and the tracking stations, which were located on the Russian landmass.
- There should be a new moon, as viewed from Earth, with the farside illuminated by the sun. Zond should arrive at the moon when it was between 24 and 28 days old.
- The parking orbit must be aligned with the plane of the moon's orbit.
- Reentry posed a real dilemma. Zond could reenter over the northern hemisphere, in full view of the tracking stations, but the long reentry corridor would bring the spaceship down over the Indian Ocean, where it would have to splash down. The cosmonauts would therefore be out of contact with the ground during this final phase, waiting for recovery ships to find them and pick them up.
- Alternatively, Zond could reenter over the Indian Ocean in the southern hemisphere, out of radio contact, but come down in the standard landing zone for Soviet cosmonauts in Kazakhstan. This offered a traditional landing on dry land, the prospect of being spotted during the descent and a quick recovery. Generally, this was the favoured approach and the one that would probably have been followed on a manned mission.

Opinions in the space programme were divided about the wisdom of splashdowns. Chief Designer Mishin was in favour, believing they presented no particular danger. Many others were against, arguing that the descent module was not very seaworthy,

The skip trajectory

was difficult to escape in the ocean and could take some time to find. They also argued against the expense involved in having a big recovery fleet at sea.

The timing for Russian around-the-moon missions around the sun–Earth–moon symmetries was therefore quite complex [14]. To meet all these requirements, there are only about six launching windows, each about three days long and a month apart, each year. There can be long periods when there are no optimum conditions. There were no optimum launch windows for Zond to the moon between January and July 1969, the climax of the moon race. The scarcity of these opportunities explains why several L-1s (e.g., Zond 4) were fired away from the moon. Although these missions caused mystery in the West, the primary Russian interest was in testing navigation, tracking and the reentry corridor. Having the moon in the sky was not absolutely necessary for these things and since it was not available anyway, they flew these missions without going around the moon.

L-1/Zond

Length	5 m
Diameter	2.7 m (base)
Span	9 m
Weight	5,680 kg
Habitable volume	3.5 m^3
Engine	One KDU-35
Fuel	AK27 and hydrazine
Thrust	425 kg
Specific impulse	276 sec

Block D

Weight	13,360 kg
Fuel	Oxygen and kerosene
Thrust	8,500 kg
Specific impulse	346
Length	5.5 m
Diameter	3.7 m

Source: Portree (1995); RKK Energiya (2001)

When the L-1 Zond was wheeled out for its first test – Cosmos 146, set for 10th March 1967 – the three-stage UR-500K Proton stood over 44 m tall and must have been a striking sight. The first two tests were called the L-1P, P for 'preliminary' indicating that a full version of Zond would not be used and that a recovery would not be attempted. The first stage would burn for 2 min with 894 tonnes of thrust. The second stage would burn for 215 sec. The third stage would place Zond or the L-1 in low-Earth orbit in a 250 sec burn. Finally, the Korolev block D fourth stage would fire 100 sec to achieve full orbit. One day after liftoff, the fourth stage, block D, would relight on the first northbound pass over the equator to send Zond out to a simulated moon [15].

Cosmos 146 was the fifth flight of Proton and the first with a block D. The block D's single 58M engine had 8.7 tonnes of thrust and burned for 600 sec, enough to accelerate the payload to 11 km/sec. In the event, block D successfully accelerated the cabin to near-escape velocity, with Cosmos 146 ending up in an elliptical high orbit reaching far out from Earth (though its ultimate path was not precisely determined). Signals and communications tests were carried out. This was an encouraging start to the L-1 programme, setting it on course for the first lunar circumnavigation by the target date of June 1967. Then the landing missions could get under way [16].

Cosmos 154 on 6th April 1967 was designed to repeat the mission to a simulated moon. This time the BOZ ignition assurance device failed during the ascent to orbit and was not in a position to control the block D stage for the simulated translunar burn, which could not now take place. This was a setback, now made worse by the crash of the related Soyuz spacecraft 20 days later and which raised questions about Zond's control and descent systems. Zond's parachute system was retested. When two such tests took place in Feodosiya in the Crimea, in June, the parachute lines snarled. Modifications took all summer. The programme underwent a thorough safety review in early September, being reviewed by an expert commission with nine working groups. Although Russia would have loved to celebrate the 50th anniversary of the revolution with a flight around the moon that November, the chances of doing so safely slipped further and further into the distance.

Working overtime, the designers and launch teams got the third L-1 Zond out to the pad by mid-September 1967. The countdown began for a launch on 28th September. The aim was to fly the L-1 Zond out to the moon and return for recovery at cosmic velocity, 11 km/sec, coming down 250 km north of Dzhezhkazgan on 4th October (or failing that in the Indian Ocean). The huge red-and-white Proton booster, weighing a record 1,028,500 kg, Zond cabin atop, tipped by a pencil spear of an escape tower, was taking with it Russia's moon hopes. It sat squat on its giant pad, shrouded by its gantry, as engineers fussed with one technical problem after another. Yet it all went wrong. One of the six engines in the first stage of Proton failed to operate when a rubber plug was dislodged into the fuel line. At 60 sec the rocket veered off course and impacted 65 km downrange, but the Zond cabin was dragged free by the escape system. The cabin was found intact the next morning, though recovering it was difficult, for the toxic burning remains of Proton were all round about.

For the anniversary of the Revolution, the Russians were left with carrying out the mission that had been intended for Soyuz that April, but now without cosmonauts on board. What happened was important for the lunar programme, but not the kind of event that would bring throngs of excited crowds out onto the streets. Cosmos 186 was first to appear, beginning a series of flights that would requalify Soyuz for manned flight once more. It went up on 27th October and was followed three days later by Cosmos 188. Using totally automatic radar, direction-finding and sounding devices, Cosmos 186 at once closed in on 188 in the manoeuvre Komarov was to have carried out in April. The rendezvous and docking manoeuvres that followed went remarkably smoothly, although the double mission was plagued with other difficulties later. At orbital insertion, 188 was only 24 km away from its companion. Cosmos 186 closed rapidly, within two-thirds of an orbit. One hour later, over the South Pacific, they clunked together to form an automatic orbiting complex and 3.5 hours later they separated. Cosmos 186 was recovered the next day and 188 was deorbited on the 2nd November (it was blown up when it came down off course). Although not visually impressive to a spectacular-weary public, it was a display of advanced robotics. It proved the feasibility of first-orbit rendezvous, the viability of Soyuz-style docking and took some of the fears out of lunar orbit rendezvous when all this would have to be done a third of a million kilometres away.

However, the elation surrounding the Cosmos 186–188 mission was followed by a disheartening experience three weeks later. The next attempt to launch Zond, the fourth, was made very early on 23rd November and was aimed at a lunar flyby and recovery. The first stage behaved perfectly, but four seconds into the second-stage burn, one of the four engines failed to reach proper thrust. The automatic control system closed down the other three engines and the emergency system was activated. The landing rockets fired prematurely during the descent and the parachute failed to detach after landing, but the scratched and battered cabin was recovered. Proton itself crashed 300 km from where it took off.

The next L-1 Zond, the fifth, got away successfully on 2nd March, 1968. This time the UR-500K Proton main stages and block D worked perfectly. Zond 4 was fired 354,000 km out to the distance of the moon, but in exactly the opposite direction to the moon, where its orbit would be minimally distorted by the moon's gravitational field. The primary purpose of the mission was to test the reentry at cosmic velocity, so going round the moon itself was not essential. Cosmonauts Vitally Sevastianov and Pavel Popovich used a relay on Zond 4 to speak to ground control in Yevpatoria, Crimea.

The Zond 4 mission was not trouble-free and the first problems developed outbound. The planned mid-course correction was aborted twice because the astro-navigation system lost its lock on the reference star. When the correction did take place, it was extremely accurate and no more corrections were required. Zond 4 was supposed to dive into the atmosphere to 45 km, before skipping out to 145 km before making its main reentry. The tracking ship off West Africa, the *Ristna*, picked up signals from Zond indicating that the skip manoeuvre had failed and that it would make a steep ballistic descent, bringing it down over the Gulf of Guinea. On the insistence of the defence minister, Dmitri Ustinov, who was afraid that it might fall into foreign hands, the spacecraft was pre-programmed to explode if it made such a descent. Accordingly, Zond 4 was blown apart 10 km over the Gulf of Guinea. Not everyone was in agreement with this extreme approach to national security. It transpired that Zond 4 was actually 2 km from dead of centre in its reentry corridor (the tolerance was 10 km), but that a sensor had failed, preventing the skip reentry.

Some consolation could be drawn from a repeat of the Cosmos link-up of the previous winter. On 15th April 1968, Cosmos 212 (the active ship) linked to Cosmos 213, this time in a record 47 min. Television showed the last 400 m of the docking manoeuvre as they aligned their wing-like panels one with another. Millions saw the separation 3 hr 50 min later over the blue void of the Pacific.

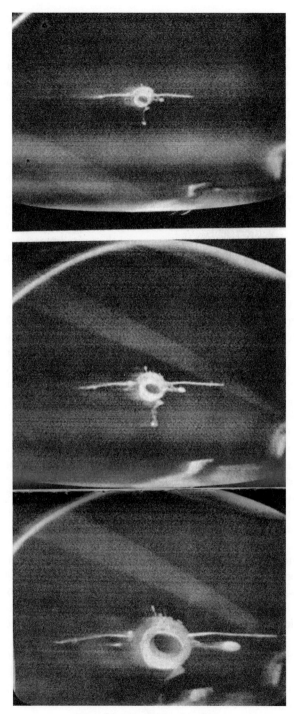

Rendezvous in Earth orbit

By the end of April 1968, the problems experienced by Zond 4 had been cured and the time was ready to try the first circumlunar flight to a 'real' moon this time. Launch took place on 23rd April. Unfortunately, 195 sec into the mission, the escape system triggered erroneously, shutting down all the Proton engines and flinging the Zond capsule clear, saving the cabin which came down 520 km away, but thereby wrecking the mission in the process. A replacement mission was planned for 22nd July, but in a bizarre pad accident in which at least one person died, block D and the L-1 toppled over onto the launch tower. Extracting the stages without causing an explosion took several nail-biting days. Further launchings were then postponed till the autumn.

The early L-1 Zond missions

10 Mar 1967	Cosmos 146
8 Apr 1967	Cosmos 154 (fail)
28 Sep 1967	Failure
23 Nov 1967	Failure
4 Mar 1968	Zond 4
23 Apr 1968	Failure

Requalification of Soyuz

27 Oct 1967	Cosmos 186
30 Oct 1967	Cosmos 188
14 Apr 1968	Cosmos 212
15 Apr 1968	Cosmos 213

THE SPACESHIP FOR ORBITING THE MOON: *THE LUNIY ORBITALNY KORABL*, LOK

The Soviet moon ship was the LOK (*Luniy Orbitalny Korabl*). Unlike the L-1 Zond, the LOK had a direct point of comparison with American hardware – the Apollo command-and-service module. Sixteen began construction, seven were completed and parts of four can still be found in museums. The LOK flew only once, on the fourth N-1 launch in November 1972, when it was destroyed, although the descent module was saved by the escape system. The traditional engineering view of the LOK is that it was a beefed-up Soyuz able to fly to the moon, but it was much more capable than that – a versatile lunar spaceship in its own right, a worthy contemporary to Apollo [17].

The descent module was the same as the normal Soyuz – but designed for a crew of two, not three; and with a thicker heat shield for the high reentry speed. The LOK weighed more, 3,050 kg, rather than 2,850 kg. The orbital module was similar to the normal Soyuz, but with different instrumentation, controls and many additional portholes for lunar orbit observations. The spacesuit for the moonwalk would be housed here, and it was from this module that the spacesuited cosmonaut would leave on his moonwalk to climb into the lunar module (LK) and begin the descent to the lunar surface. The orbital module had a large hatch, 90 cm, sufficiently wide to permit the cosmonaut to exit in the *Kretchet* lunar suit. The orbital module had a control unit for masterminding the link-up in lunar orbit after the landing and a forward-looking porthole. Rendezvous and docking would be controlled from there, not from the descent module.

Compared with Soyuz, it had a much larger skirt at the base, an additional small forward module and a docking system at the front, called *Kontakt*. A series of antennae and helices were used to zone in on the returning landing module, the LK, for rendezvous and docking. The LOK's probe, called *Aktiv*, would penetrate an aluminium plate on the top of the LK. It had 108 recessed honeycomb hexagons on a plate 100 cm across and entry to only one of these would be sufficient to achieve a firm capture.

The most visible differences from Soyuz were in the instrument-and-propulsion module at the rear and the small extra module at the front. The 800 kg front module contained six fuel tanks, each with 300 kg of UDMH, four engines for attitude control in lunar orbit, an orientation engine and the *Kontakt* docking unit. On Apollo, there was a small conical docking unit on the front of the command module, but the other elements were made an integral part of the service module. For rendezvous, the LOK closed in on the LK in lunar orbit, the flight engineer peering through the forward-looking porthole, using television and handling an adjacent control panel. The front module of the LOK had four attitude control thruster units, each with two main nozzles and two small ones. The engine system was made by the Arsenal Design Bureau in Leningrad.

At the rear, the LOK carried two propulsion sets. The biggest was the main engine for the return to Earth, the equivalent of the Service Propulsion System of Apollo. The LOK's engine had a thrust of 3,388 kg and a specific impulse of 314 and its primary purpose was to make the trans-Earth injection burn out of lunar orbit. The engine, called the S5.51, was built by the Isayev design bureau. The LOK also carried the standard Soyuz engine, to be used as a rendezvous motor, with a thrust of 417 kg, a specific impulse of 296 and capable of 35 restarts. The LOK carried 2,032 kg of nitrogen tetroxide and 1,120 kg of UMDH. The LOK was the first Soviet spacecraft to carry the fuel cells pioneered by the Americans in the Gemini programme: 20 Volna cells, weight 70 kg, able to supply 1.5 kW for ten days. They were made by the Ural Electrochemical Enterprise. The only other Soviet spaceship to carry fuel cells was the *Buran* space shuttle in 1988. The rear section carried radiator shutters to shed heat. At the junction with the descent module were star trackers.

LOK's arrival in lunar orbit followed a different procedure from Apollo. The mid-course manoeuvre and lunar orbit insertion were done by block D, not by the LOK's main engine. Block D would again be used to lower the orbit of the LOK and LK over the lunar surface to its final orbit dipping to 16 km and, finally, for all but the final part of the powered descent of the LK. On Apollo, the Service Propulsion System carried out the mid-course correction moonbound, lunar orbit insertion and lunar orbit corrections.

With the LK down on the surface, the profile of the LOK now closely approximated that of the Apollo command-and-service module. The LOK would orbit the moon, a sole cosmonaut flight engineer aboard, like the single astronaut on the Apollo. For half of each orbit, it would be around the farside of the moon, out of contact with the Earth. Once the LK blasted off from the lunar surface, it was the task of the LOK to locate the rising LK, close in and dock. The *Kontakt* system was designed in such a way that a simple contact would join the spacecraft together, so there was no question of hard and soft dockings. Unlike Apollo, the LK cosmonaut would transfer externally back to the LOK by spacewalk. The LK would, like the American LM, then be jettisoned. The LOK would then make the crucial burn out of lunar orbit, make the three day coast back to Earth, carry out two mid-course corrections (one at mid-point, one just before reentry) and then make a Zond-type skip reentry.

LOK

Weight (at LOI)	9,850 kg
(at TEI)	7,530 kg
(on return)	2,804 kg
Length	10.06 m
Diameter	2.93 m
Habitable volume	9 m^3
Crew	2
Max. flight time (days)	13
Descent module length	2.19 m
diameter	2.2 m

Source: RKK Energiya (2001)

THE SPACESHIP TO LAND ON THE MOON: THE *LUNIY KORABL*, LK

The descent of the Soviet lunar lander, called the LK (*Luniy Korabl*), to the lunar surface would be a steep one. The final lunar orbit would be 16 km by 85 km, the same as the final orbit of the later Ye-8–5 lunar sample return missions. Block D would fire at the 16 km perilune, bringing the LK to between 2 km altitude (maximum) and 500 m (minimum), ideally 1,500 m. If all went well, the LK pilot would set the LK down about 25 sec thereafter, but not more than a minute later. The LK would descend to 110 m, when it would hover: then the cosmonaut would take over for the landing. The instructors told the cosmonauts that at 110 m, they had three seconds to select a landing site, or return to orbit ('as if returning at this stage was an option', snorted Leonov). The standing cosmonaut, watching through his large, forward-looking window, would guide the LK lander with a control stick for attitude and rate of descent.

The engine, called block E, was designed by the Mikhail Yangel OKB-586 in Dnepropetrovsk. It was a well-equipped propulsion set. The LK module had:

- One 11D411 RD-858 main engine weighing 53 kg with a single nozzle with a specific impulse of 315 sec, chamber pressure of 80 atmospheres and duration of 470 sec.
- A 11D412 RD-859 57 kg backup engine with two nozzles.
- Four vernier engines.
- Two 40 kg thrusters for yaw.
- Two 40 kg thrusters for pitch.
- Four 10 kg thrusters for roll.

The descent and take-off engine was a throttlable, single-nozzle, 2.5-tonne rocket burning nitrogen tetroxide and UDMH. It could be throttled between 860 kg thrust and 2,000 kg. The engine held 1.58 tonnes of nitric acid and 810 kg of UDMH. The engine had four verniers to maintain stability. For attitude control during the nerve-wracking descent to the moon, eight low-thrust engines designed by the Stepanov Aviation Bureau fed off a common 100 kg propellant reserve. The system was both safe – it ran off two independent circuits – and sensitive, for thrust impulses could last as little as nine milliseconds. To land the LK, the cosmonaut had a computer-assisted set of controls, the first carried on a Soviet manned spacecraft. The S-330 computer was a sophisticated digital machine, linking the cosmonaut's commands to the lander's gyroscopes, gyrostabilized platform and radio locator, with three independent channels working in parallel [18]. Four upward-firing solid rockets would ignite on landing, to press the LK onto the surface. The lander was designed to take a slope of 20°.

   The LK was different from the Apollo lunar module (LM) in a number of important respects. These were a function of the much poorer lifting power of the N-1 rocket. First, it was much smaller, being only 5.5 m tall and weighing 5 tonnes (the LM was, by contrast, 7 m tall and weighed 16 tonnes). It had room for only one cosmonaut standing and the lower stage would have no room for the extensive range of scientific instruments carried by Apollo. Second, the LK had a single 2,050 kg thrust main engine which was used for both descent and take-off (Apollo's LM had a descent motor and a separate one for the small upper stage). Like the LM, the LK would use the descent stage as a take-off frame. The LK was designed for independent flight of 72 hours and up to 48 hours on the lunar surface. The LK was a minimalist approach to a lunar landing. Although the method of landing on and take-off from the moon was broadly similar, there were some important differences:

- The American LM descent engine carried out the entire 12 min descent from PDI (powered descent initiation) to touchdown. By contrast, block D provided most of the thrust of the descent of the Soviet LK. Block D was dropped around 1,500 m above the surface and the LK's descent stage took over for the final part.
- The American LM had two motors, one for descent and one for ascent. By contrast, the Russian LK had just one motor, which was used for descent and ascent.

What would the LOK–LK mission have been like? It would begin with the launching, from Baikonour Cosmodrome, of two cosmonauts on the N-1 rocket. The three stages

The LK

of the N-1 rocket would burn until the lunar stack was safely in an Earth orbit of 51.6°, 200 km. At the end of the first parking orbit, the fourth stage, block G, would fire for translunar injection. This block would then separate.

Unlike Apollo, there would be no transposition, docking and ejection of the lunar module. This would remain behind the command ship, the LOK, as they headed moonward. On the way to the moon, the fifth stage, block D, would fire for a translunar correction. Three days into the mission, block D would fire the stack into lunar orbit. The descent from lunar orbit would again be different from Apollo. First, a lone cosmonaut would enter the lunar module, the LK. Because there was no internal hatch, the cosmonaut would exit the hatch and climb down the side of the LOK along a pole before entering the access hatch. This would take place against the backdrop of the moon's surface below and the spectacle would be stunning. Once on board the LK, the cosmonaut would then separate his lunar module and block D from the LOK mother ship. Here would come a fresh difference. The powered descent

LK hatch

LK inside

LK ladder

LK window

burn would be done by block D. It would be jettisoned a mere 1,500 m above the lunar surface, leaving the LK's main engine to complete the descent to the lunar surface. This would be the same engine used for take-off.

Hover time was much tighter on the Russian LK than the American LM. The Russians had about a minute to find the landing site and put the spacecraft down. The pilot could, of course, use more than 1 min, since it was the same engine used for the ascent, but this would eat into the thrust required for ascent. The LM had a longer hover time, about 2 min. By the end of the 2 min, the LM would be out of fuel and the mission would have to abort. Below a certain altitude, the period of time for firing the ascent stage would be longer than the time taken to fall to the surface, so the LM would crash (this was called 'dead man's handle'). All but one of the Apollos were sufficiently well targeted not to present a problem. The most difficult landing was the first, Apollo 11, which landed with only 19 sec of fuel to spare. 'Dead man's handle' did not operate on the LK, since the engine used for the ascent was already firing. Arguably, it was safer. The LK lunar lander, like Apollo, had four legs. The first Soviet moon landing would have been shorter than that of Apollo 11, without a sleep period.

Once on the surface, the sole cosmonaut would carry out a spacewalk. We do not know how long the first lunar stay was planned. A moonwalk duration of four hours has been suggested, so the surface stay time would have to be long enough to report back after landing, prepare for the moonwalk, carry it out, return and prepare for take-off and rendezvous.

After several hours on the surface, the cosmonaut would lift off from the moon in the upper stage of the LK, and conduct the type of rendezvous pattern tested by Cosmos 186–188, 212–3 and Soyuz 2–3 and 4–5 in which the LOK orbiter performed the active role. A backup two-nozzle engine was also available should the motor fail to light for the critical liftoff from the moon. On liftoff, the backup engine was actually fired simultaneously with the main engine, but turned off if the main engine lit up. The LK had five chemical batteries, three on the descent stage, two on the ascent. Cabin pressure was oxygen/nitrogen at 560 mm.

The return-to-Earth profile was quite like Apollo. The LK would lift off from the lunar surface, using the landing frame as a launching pad, like the American LM. The LK would link up with the LOK in lunar orbit and the cosmonaut would transfer to the LOK, though this would be by an external spacewalk (indeed, it would be his third that day). The LK would be dropped, and then the LOK would fire its main engine for trans-Earth injection. There would be a quiet coast Earthward, followed by a high-speed skip reentry over the Indian Ocean and a soft landing in Kazakhstan.

The LOK and L-1 spacecraft were expected to return to Earth in the standard recovery zone in Kazakhstan. Here, the Russians had extensive experience of the Air Force recovering spacecraft using helicopters, trucks, amphibious vehicles, adapted troop carriers and other vehicles able to traverse the flat steppeland. This experience had been built up during the Korabl Sputnik missions and the Vostok series and consolidated as the military photoreconnaissance Zenit series began making regular missions. The real problem was if the L-1 or LOK came down outside Soviet territory, either by choice or if the skip return failed and a ballistic path was followed instead. The Indian Ocean was the most likely maritime landing point. Here, in a decree issued on 21st December 1966, the Soviet Navy was made responsible for Indian Ocean recoveries. For Indian Ocean recoveries, ten naval and maritime research ships were involved, supplemented by three ship-borne helicopters, spread out at 300 km points along the ocean.

The LK

Weight	5,500 kg
Height	5.2 m
Diameter ascent stage	3 m
Span, descent stage	4.5 m
Habitable volume	4 m^3
Hover time	1 min
Weight, ascent stage	2,250 kg
Weight, descent stage	2,250 kg
Crew	1
Length of legs	6.3 m

Were Soviet computers up to the job? The Apollo 11 American lunar landing nearly aborted when the lunar module's computer overloaded and flashed alarms in the LM cabin. The Apollo computers, though the most sophisticated of their day, would be regarded as laughably primitive nowadays. They were bulky, crude and had limited memory, but they played an important part in getting Apollo to the moon and back again. The popular assumption is that Soviet computers during the moon race lagged far behind American ones. This does not seem to be the case now. The Soviet Union had a long tradition in advanced mathematics and developed, in the late 1950s, its own silicon valley, partly assisted by two exfiltrated American electrical engineers, communists and friends of the Rosenbergs, Alfred Sarant and Joel Barr [19]. Taking on fresh names, Philip Staros and Josef Berg, they built up Special Design Bureau 2 (*Spetsealnoye Konstruktorskoye Buro 2*, SKB 2) which developed microcomputers for the Soviet aviation industry, military and space programmes. This included the *Argon* computer used on Zond. During the 1960s, SKB 2 developed a series of small, lightweight, sophisticated computers, from laptops to navigational devices to big calculating computers. Just because Soviet computers followed a different development path from the West did not mean that they were inferior, for they were not. The ability of the USSR to achieve automated rendezvous and docking in space (1967) went unmatched in the West until 1998 when the Japanese satellites *Hikoboshi* and *Orihime* met in orbit.

RUSSIA: THE MOONWALK

A special spacesuit was required for the moonwalk. The design requirements for a moonsuit were much tougher than for normal spacewalking, for they required:

- Long duration, so as to make possible a proper programme of lunar surface exploration.
- Spare duration, in the case of difficulty in returning to the LK.
- Tough soles and boots for the lunar surface.
- Durability, so it would not tear if the cosmonaut fell onto the lunar surface.

Russian spacesuits went back to Air Force pressure suits and balloon flights in the 1930s [20]. For the first manned orbital missions, a bright orange pressure suit was developed. The first suit for spacewalking was developed in 1963, called the *Berkut*. This was used by Alexei Leonov for the first ever spacewalk in March 1965. After this, in anticipation of similar manoeuvres on moon flights, requirements were issued for the testing of a spacesuit suitable for the external transfer between orbiting spacecraft. These refinements were tested by cosmonauts Yevgeni Khrunov and Alexei Yeliseyev in January 1969 and this suit was called the *Yastreb*. It was the first purely autonomous spacesuit, without an air supply from the cabin, using a closed-loop life support system. In the course of a 1 hr spacewalk they transferred from Soyuz 5 to Soyuz 4, using a backpack strapped to their legs (the hatches were too wide for the packs to go on their backs).

For the lunar surface spacewalk, a special spacesuit was developed [21]. Chief Designer Vasili Mishin laid down the requirement for a special, semi-rigid spacesuit for the moonwalk, although the contrary view was expressed that it would have been easier to develop a version of the *Berkut* spacesuit used by Alexei Leonov for Voskhod 2. Design began in 1966. The suit was called *Kretchet*, though to be more precise *Kretchet* was the experimental model and *Kretchet 94* the final operational version. Responsibility for the spacesuit fell to the Zvezda bureau of Gai Severin, the company which had made the previous suits. The design was finally agreed on 19th March 1968. During this period, the Zvezda bureau also designed and built a traditional soft suit called *Oriol*, but the higher performing *Kretchet* appears to have been the favourite all along.

Cosmonaut on the lunar surface

Unlike the American suit, or the earlier Russian suits, which were donned piece by piece, the Russian suit was a semi-rigid, single-piece design that the cosmonaut climbed into through a door at the back. This was a radical departure, making the *Kretchet* virtually a self-contained spaceship in itself. The idea was not a new one: it had been sketched in detail by Konstantin Tsiolkovsky in 1920 in his science fiction novel *Beyond the planet Earth*. An important advantage of the suit was its one-size-fits-all approach: cosmonauts inside it were able to adjust its dimensions according to their size. By contrast, the American moonsuits were individually tailored. The *Kretchet* could be donned – or entered – quickly and it did not take up any more room in the cabin than a traditional suit.

The mission commander first donned the *Kretchet* in the LOK, before using it to spacewalk over to the LK for the descent to the lunar surface. The *Kretchet* was designed to work for up to 52 hours, up to six hours at a time and enable the cosmonaut to venture as far as 5 km from the lander. Surface time was estimated at four hours, with 1.5 hours contingency and a half-hour red line emergency (in fact, the designers provided up to ten hours at a time). The suit originally weighed 105 kg on Earth, about a fifth of that on the moon. The moonwalker monitored and controlled the functions of the spacesuit by a fold-down panel console on the front. The suit was designed to be tough, with ten layers of protection.

The *Kretchet* was designed so that it could be used independently or hooked up to the cabin of the LK and replenished from the LK's own atmospheric supply. The 52 hr requirement was set down with a view to the suit keeping the cosmonaut alive

Orlan, descendant of the *Kretchet*

during takeoff and rendezvous should the LK fail to repressurize after the spacewalk. A bizarre feature of the *Kretchet* was that the designers put around it a kind of hula-hoop ring. The purpose was to ensure that if a cosmonaut fell over, something they worried about, he could use the ring to bounce back up. The Americans had no such system, probably because one astronaut could help his colleague pick himself up if he fell. Later, television viewers saw the later Apollo astronauts fall over many times, doing themselves little evident harm and presenting little danger.

A less rigid version of *Kretchet* was devised for ordinary spacewalking. This was called *Orlan*. This followed the same principles but did not carry the hoop, the heavier moonwalk boots nor as large air tanks (2 hr rather than 5 hr). Orlan relied on power supplied from the spaceship, rather than internal systems. It was lighter (59 kg), had only five layers of protection and no waste removal system. This suit would be worn by the flight engineer on board the LOK. In the event of the commander experiencing difficulty going down to the LK or returning therefrom, the flight engineer could venture out in the *Orlan* to retrieve him.

Twenty-five *Kretchet* suits were built for testing and training over 1968–71. They were given to the cosmonaut squad for testing. They were put through thermal and vacuum tests in a simulated moon park in the Zagorsk rocket engine test facility. The operation of the suit was checked in a Tupolev 104 aircraft, as were tools for use on the lunar surface. Work on the original *Kretchet* suit was suspended in 1972 and then, on orders from Valentin Glushko, terminated on 24th June 1974. Nine were in production at the time. In the course of 1972–4, when work was focused on the N1-L3M programme, the Zvezda design bureau began work on a more advanced version of the *Kretchet* in the light of the much more ambitious surface expeditions envisaged under the N1-L3M.

Unlike some of the hardware from the lunar landing programme, the Soviet moonsuit story has a happy ending. *Kretchet–Orlan* was a successful design and subsequently used on the Salyut space stations from 1977 onward and on Mir thereafter. A new version, *Orlan M*, was introduced on the space station Mir and later became the Russian suit used on the International Space Station. It is reckoned to be one of the best spacesuits ever made. The experience gained in developing *Oriol* was also put to use in the development of the subsequent *Sokol* Soyuz cabin suit. So almost 40 years later, the successors of the Russian moonsuits are still in good use.

The Russian moon suit, *Kretchet*

Weight	105 kg
Duration (total)	52 hr
Surface	6 hr+

We know little of what the Soviet cosmonaut would have done on the lunar surface. Our only account comes from I.B. Afanasayev's monograph *Unknown spacecraft* (1991), one of the early histories of the Soviet moon programme. This is what he says:

The operations on the moon would consist in planting the USSR state flag, deploying the scientific instruments, collection of the lunar soil samples and photographing the terrain, as well as conducting television reportage from the lunar surface. The arsenal of scientific instruments at the disposal of the Soviet cosmonaut would be extremely restricted by the low weight of cargoes that the LK could carry.

According to Mishin, the lander would have two deployable antennae, two sets of surface experiments and the possibility of a small rover [22]. The best information suggests that the moonwalk would take four hours, not more than 500 m from the cabin, but that the cosmonaut should be able to walk up to 5 km to a reserve LK which, in at least one plan, would be landed nearby. A shorter moonwalk time was also considered, using one of the earlier types of spacesuits (*Yastreb*), but Mishin held out for a full-length moonwalk using the *Kretchet*. No decision was taken, but granted that *Kretchet* was available it seems likely that a full-length moonwalk would have been undertaken.

Collecting the soil sample would have been the first task, as it was on Apollo, so that if the moonwalk had to be aborted, the cosmonaut would at least not return empty-handed. Just as President Nixon made a phone call from the White House to the Apollo 11 astronauts, Leonid Brezhnev would certainly have sent a similar message (he did during the Apollo–Soyuz mission in 1975).

The LK was designed to carry a three-piece surface package. Details are sparse, but they have been assembled by the expert on Soviet space science, Andy Salmon. The main package comprised two seismometers, each with four low-gain aerials, shaped a little like a landmine covered in thermal blankets, to be positioned equidistant from opposite sides of the LK. As was the case with the American LM, they were carried in the lower stage of the LK and then lifted to their chosen locations by the cosmonaut. The third item was a small crawler or micro-rover, to be deployed at the end of a cable supplying power and communications from the LK. Such a rover, called PrOP-M, was built for Mars missions at this time. Presumably, the rover would be remote-controlled from Earth once the cosmonauts had left the moon and then manoeuvred slowly across the lunar surface. The design has a number of similarities to cabled crawlers carried to the Red Planet by Mars 3 in 1971. Designer of the LK surface package is understood to have been Alexander Gurschikin of the Academy of Sciences.

TRACKING

To support the manned lunar effort, a fleet of maritime communication ships was constructed for the period when the manned lunar-bound or home-bound spacecraft would be out of direct line with the tracking stations in the Soviet Union. Initially, some merchant ships were converted to carry tracking equipment: the *Ristna* and *Bezhitsa*. Then some newly converted merchant ships were introduced: *Kegostrov*, *Nevel*, *Morzhovets* and *Borovichi*. Some small ground stations were set up in friendly states: in Chad, Cuba, Guinea, Mali and the United Arab Emirates.

Tracking ship *Cosmonaut Yuri Gagarin*

The big step forward was in May 1967, when a new class of large tracking ship was introduced, starting when the *Vladimir Komarov* was spotted making its way down the English Channel for its shake-down cruise. The *Cosmonaut Vladimir Komarov* displaced over 17,000 tonnes and had a crew of several hundred with large radar domes and antennae. Large though it was compared with its predecessors, it was small compared with the much larger tracking ships that followed: the *Cosmonaut Yuri Gagarin* (45,000 tonnes), which became the flagship and the *Academician Sergei Korolev* (21,250 tonnes). For Western observers, knowledge of the whereabouts of the tracking fleet was important in predicting Soviet lunar or manned missions. If the tracking ships were at sea, then missions could be expected (this became equally true of the Chinese manned space programme 30 years later).

THE COSMONAUT SQUAD

So much for the hardware. What about the people who would fly to the moon? The selection of cosmonauts for the moon programme went through a number of phases:

- The selection of cosmonauts for the general moon programme.
- The division of this group into candidates to train for the around-the-moon flight (L-1) and the moon landing itself (the LOK and the LK). Some cosmonauts belonged to both.
- Selection of cosmonauts for the first around-the-moon and landing missions.
- Decline and disbandment of the group. All then returned to mainstream missions.

Although no cosmonaut ever did make the trip around the moon or to its surface, we know with a fair degree of certainty who would have made these voyages [23].

Russia drew on its existing teams of cosmonauts for its moon missions. By 1970, the Soviet Union had selected a number of cosmonaut groups. Essentially, Soviet cosmonaut selection was divided into three streams: Air Force pilots and military engineers, who commanded missions; flight engineers, civilians mainly drawn from the design bureaux that made the spacecraft; and specialists, like doctors, selected for specific missions. By the time of the moon programme, the following groups of pilots had been selected:

- Twenty young Air Force pilots for the first manned spaceflights (1960).
- Five young women to make the first flight by a woman into space (1962).
- Fifteen older Air Force pilots and military engineers (1963) (two more joined the group later).
- Twenty young Air Force pilots and military engineers (1965), later called 'the Young Guards'.

The following groups of civilian engineers had been selected:

- Two engineers, one of whom would fly on the first multi-manned Voskhod flight (1964).
- Six engineers from OKB-1 (1966), with three more joining the following year.
- Three more civilian engineers (1969).

The following specialists were also selected:

- Two doctors (1964).
- Four Academy of Sciences cosmonauts (1967).

Many more cosmonauts were selected subsequently, but too late for the prospective moon missions and they are not considered here (the much later N1-L3M plan never got so far as to merit the selection of cosmonauts). Of the groups above, two were not relevant to the moon programme. The women's group was selected for the first flight of a woman in space, eventually made by Valentina Tereshkova in 1963. Although there was a number of discussions about further missions by women, none came to fruition and none were ever considered for a moon mission. The group was disbanded in 1969. The two doctors likewise were never considered for the moon mission.

For its moon mission, Russia theoretically had available up to 74 cosmonauts. In reality, the total number available was much smaller, for some had retired or gone on to other work. A small number died during accidents. By far the largest cause for the reduction of numbers was people exiting due to failing medical tests, sometimes caused by the rigorousness of the training régime. A small number was also dismissed for indiscipline.

Early Soviet cosmonauts, Sochi, 1961

Those chosen for the moon mission were inevitably likely to be drawn from the most experienced members of the groups, especially those who had flown in space before. In more detail, the following is the pool from which they were drawn:

1960 first Air Force pilot selection (20): Ivan Anikeyev, Pavel Belyayev, Valentin Bondarenko, Valeri Bykovsky, Valentin Filateyev, Yuri Gagarin, Viktor Gorbatko, Anatoli Kartashov, Yevgeni Khrunov, Vladimir Komarov, Alexei Leonov, Grigori Nelyubov, Andrian Nikolayev, Pavel Popovich, Mars Rafikov, Georgi Shonin, Gherman Titov, Valentin Varlamov, Boris Volynov, Dmitri Zaikin.

This was the first, original and most famous group of cosmonauts. These were the equivalent of the Mercury seven, selected in April 1959 for the first American mission into space and immortalized in the film *The right stuff*. Russia's right stuff comprised young Air Force pilots recruited in 1959–1960. Compared with the American group, they were much younger (24 to 35, but mainly at the younger end) and had much fewer flying hours. Gherman Titov, the second Russian to orbit the Earth, was only 25 years old when he made his mission. Yuri Gagarin, the first man in space, had only 230 flying hours to his credit when he joined the cosmonaut squad (prospective Americans must have a minimum of 1,500). Like the Americans, the Russians put an emphasis on young, tough, fit men in perfect health who could react quickly to difficult situations. Young Air Force pilots, disciplined by military service, were considered to provide the best possible background for the early space missions. China selected a similar type of person for its first *yuhangyuan* group (1970) and its second one many years later (1996).

Cosmonauts from the moon flights were most likely to be drawn from this group. By autumn 1968, eight members of the group had flown, in this order: Yuri Gagarin, Gherman Titov, Andrian Nikolayev, Pavel Popovich, Valeri Bykovsky, Vladimir Komarov, Pavel Belyayev and Alexei Leonov. There was a high rate of attrition from this group and eight of the group never flew in space at all because of problems,

Chief designer Valentin Glushko with cosmonauts

accidents and even dismissals due to indiscipline. Two died during the moon pro-gramme (Vladimir Komarov and Yuri Gagarin).

The first 20 cosmonauts were recruited through the Institute for Aviation Med-icine with a view to one of them making the first manned flight into space. A centre for the training of cosmonauts was approved in January 1960, called the Centre for Cosmonaut Training, TsPK, the title it still uses. General Kamanin was appointed director of the squad, a position he held until 1971. The first cosmonauts arrived in February 1960, the rest the following month, and work formally began with the first lecture at 9 a.m. on the morning of 14th March 1960. Originally, TsPK was located in an office building belonging to the MV Frunze airfield on Leninsky Prospekt, but in June 1960 the centre moved out to a greenfield location. This was a 310 ha site in birch forest, now known as Star Town (sometimes Star City), 30 km to the northeast of Moscow. A secret location until the 1970s, it was to become the most international space training centre in the world by the 1990s.

Star Town's weather crosses extremes, ranging from +30°C in high summer to −30°C in the depths of winter. A central focus of Star Town is the man-made lake, which freezes over in winter. Around it may be found accommodation for the cosmonauts and workers at Star Town, comprising 15-floor blocks. Transport is mainly by rail via the nearby Tsiolkovsky railway station or by minibus from Moscow [24]. Star Town comprises accommodation, a museum, nursery, school, health and sports centres and an hotel (called *Orbita*). It is a closed, guarded, walled town, though

Yuri Gagarin at home

entry is now much easier than in its early days. In the central area may be found, within a further walled area, the cosmonaut training centre: simulators, centrifuge, hydrolab (water tank to test spacewalks), planetarium and running track.

The high attrition rate among the first group of cosmonauts meant that a second main group should be selected and, accordingly, a new group of pilots and military engineers was chosen on 11th January 1963.

1963, second Air Force pilot selection (17): Georgi Dobrovolski, Anatoli Filipchenko, Alexei Gubarev, Anatoli Kuklin, Vladimir Shatalov, Lev Vorobyov, Yuri Artyukin, Edouard Buinovski, Lev Demin, Vladislav Gulyayev, Pyotr Kolodin, Edouard Kugno, Alexander Matinchenko, Anatoli Voronov, Vitally Zholobov, Georgi Beregovoi, Vasili Lazarev.

This group was selected for the flights of the Soyuz complex. There was an important change in direction in recruitment. There had been concerns over the individualistic bent of some members of the first squad. The cosmonaut selectors now wanted to go for slightly older, more mature pilots who would be less likely to cause discipline problems. Graduation from an institute was a requirement, ruling out young pilots straight out of school. There was more emphasis on education and flying experience, less on physical perfection. Military engineers were included for the first time. This was to prove quite a successful selection group, for many went on to become eminent and reliable cosmonaut pilots in the 1970s. One exception was the unfortunate Eduard Kugno. Strange though it might seem to Westerners, membership of the Communist Party was not a prerequisite for selection to the cosmonaut squad (designer and cosmonaut Konstantin Feoktistov was a famous non-joiner). Kugno went a stage further and when asked why he had not joined, he said he would never join a party of 'swindlers and lickspittles'. He was promptly dismissed for 'ideological and moral instability'.

Although by this stage, the cosmonaut squad consisted of over 30 members, the multiplicity of manned programmes under way suggested the need for another round of recruitment. In early 1965, the call went out for more candidates and 20 were selected from the 600 who applied. They were a mixture of pilots and engineers, with one military doctor (Degtyaryov). For this group, the Russians went back to younger pilots in their 20s, but this time making sure that they had stable psychological backgrounds.

> *1965, third Air Force selection (22) ('the Young Guards')*: Leonov Kizim, Pyotr Klimuk, Alexander Kramarenko, Alexander Petrushenko, Gennadiy Sarafanov, Vasili Shcheglov, Ansar Sharafutdinov, Alexander Skvortsov, Valeri Voloshin, Oleg Yakovlev, Vyacheslav Zudov, Boris Belousov, Vladimir Degtyaryov, Anatoli Fyorodov, Yuri Glazhkov, Vitally Grishenko, Yevgeni Khludeyev, Gennadiy Kolesnikov, Mikhail Lisun, Vladimir Preobrazhenski, Valeri Rozhdestvensky, Edouard Stepanov.

This group was not formed with moon missions in mind at all, but with a view to undertaking, after a lengthy period of training, a range of missions some time in the future. This explains the decision to go for a younger age group. They were accordingly called 'the Young Guards'. This group did, in the course of time, provide a number of pilots for space station missions in the 1970s, but it also suffered high attrition rates.

Originally, there was a broadly accepted view that 'right stuff' cosmonauts must be drawn from the military. The Russians began to recruit cosmonauts from further afield much sooner than the Americans. With the first three-man spaceship in 1964, the Voskhod, there were seats which did not need to be filled by military cosmonaut pilots. Sergei Korolev established the principle that engineers and specialists should also be regular participants on Soviet spaceflights and for the Voskhod mission awarded one seat to a designer, the other to a doctor (Boris Yegorov was the lucky man). For the civilian engineer group, two were selected, of whom one would fly, N-1

Yuri Gagarin with Valentina Tereshkova

and Soyuz complex designer Konstantin Feoktistov. He was now a senior designer in Korolev's own OKB-1 and had been involved in the design departments from the mid-1960s. When the opportunity came to fly passengers on Voskhod, he leapt at the chance.

1964 civilian engineers (2): Konstantin Feoktistov, Georgi Katys.

Konstantin Feoktistov was drawn from OKB-1 and Georgi Katys from the Academy of Sciences. This was not intended as a cosmonaut group as such, but as a selection that would train for one mission only and then return to normal duties. Feoktistov did not go back quietly, but pressed persistently but unsuccessfully to get further missions. Medical tests went against him. He continued to offer his opinions on spaceflight history and contemporary issues into the 1990s.

The first substantial group of civilian engineers was recruited by Vasili Mishin on 23rd May 1966. They did not report for training until September and some of those listed joined the group even later.

Sergei Korolev with Konstantin Feoktistov

1966 civilian engineers (10): Gennadiy Dolgopolov, Georgi Grechko, Valeri Kubasov, Oleg Makarov, Vladislav Volkov, Alexei Yeliseyev, Vladimir Bugrov, Nikolai Rukhavishnikov, Vitally Sevastianov, Sergei Anokhin (instructor cosmonaut).

Sergei Anokhin was a leading test pilot and put in charge of the group. All were drawn from OKB-1. Despite the worries of the military, there had been no disasters arising from the flying of civilians on Voskhod. The successful flight of Konstantin Feoktistov set the trend for the permanent division of cosmonaut selection into two streams: the military and civilian (with the further category of specialist). Later, it became standard for Russian spacecrews to comprise a military commander and civilian flight engineer. Because of its preeminence in the manned spaceflight programme, almost all the civilians came to be drawn from OKB-1, but in the course of time small numbers were also taken from the other design bureaux and from the Institute of Bio Medical Problems. The rationale behind the civilian selections was that those who knew most about the spaceships were those who designed them. They were the best people to fix them if they went wrong. By contrast, there was no equivalent tradition in the American space programme.

A larger selection of specialists was made several years later: four scientists.

1967 Academy of Sciences (4): Rudolf Gulyayev, Ordinard Kolomitsev, Mars Fatkullin, Valentin Yershov.

This was the first selection of scientists in the Soviet space programme [25] and was an initiative of the President of the Academy of Sciences, Mstislav Keldysh. The United

Cosmonaut Valentin Yershov, lunar navigator

States had also begun scientist selections at around this time. Eighteen scientists applied for this selection, four reaching the final selection on 22nd May 1967. Gulyayev, Kolomitsev and Fatkullin came from the Institute for Terrestrial Magnetism, Ionosphere and Radio Wave Propagation, while Yershov came from the Institute for Applied Mathematics. Kolomitsev was a true explorer, having spent over four years at the Soviet Antarctic southern magnetic pole Vostok base. The first three hoped to get assignments on Earth-orbiting missions, but Yershov was chosen with the upcoming lunar missions in mind where he would assist as navigator. He had an unhappy background, for his father, a police officer, had been killed by the NKVD secret police. Yershov, born 1928, had developed surface-to-air missiles before joining Keldysh's Institute for Applied Mathematics in 1956. There he specialized in spacecraft navigation, working on the development of the autonomous navigation system of the L-1 Zond. Yershov even developed a theorem of measurement named partly after him, the Elwing–Yershov theorem.

Finally, a group of three engineers was selected in 1969 and they were the last group whose members entered lunar selections.

1969 civilian engineers (3): Vladimir Fortushny, Viktor Patsayev, Valeri Yazdovsky.

The last two were drawn from OKB-1, Vladimir Fortushny from the Paton Institute of Welding in the Ukraine. Fortushny was selected with a view to a welding-in-space mission, not for lunar flights (the mission was flown as Soyuz 6 in 1969, but by another cosmonaut, Valeri Kubasov). So this was the pool from which the lunar missions would be drawn.

RUSSIA'S MOON TEAM: SELECTION FOR AROUND-THE-MOON AND LANDING MISSIONS

The selection of cosmonauts into two groups reflected the two streams of the moon programme: around the moon (L-1) and landings (LOK and LK). The first selection took place on 2nd September 1966, when an L-1 Zond group was established:

Commanders: Georgi Beregovoi, Valeri Bykovsky, Yuri Gagarin, Yevgeni Khrunov, Alexei Leonov, Vladimir Komarov, Andrian Nikolayev, Vladimir Shatalov and Boris Volynov.

Flight engineers: Georgi Grechko, Valeri Kubasov, Oleg Makarov, Vladislav Volkov, Alexei Yeliseyev.

This was modified 18th January 1967 and made smaller on account of the upcoming Soyuz 1/2 mission:

Commanders: Pyotr Klimuk, Alexei Leonov, Pavel Popovich, Valeri Voloshin, Boris Volynov, Yuri Artyukhin.

Engineers: Georgi Grechko, Oleg Makarov, Nikolai Rukhavishnikov, Vitally Sevastianov, Anatoli Voronov.

This officially marked the start of mission training for the lunar programme. Despite the selection, getting training under way was another matter. The first simulator for the L-1, called *Volchok*, did not arrive until a year later, in January 1968. It was built by the M.M. Gromov Flight Research Institute and installed at the Air Force Institute for Space Medicine. The main function of the simulator was to enable the training group to practise high-speed ballistic and skip reentries into the Earth's atmosphere, which was considered the point of greatest difficulty and danger. Versions of the simulator were developed for the LK (*Luch*), the descent module (*Saturn*) and for practising rendezvous (*Uranus* and *Orion*). The group did no fewer than 70 simulated returns from the moon and, according to Alexei Leonov, learned to land the simulator back on Earth with an accuracy of 1,000 m. Not only that, but they practised the reentry manoeuvre in the 3,000-tonne centrifuge in Star Town. At one stage, Alexei Leonov was subjected to 14 G, causing haemorrhages on those parts of his body that were most severely compressed.

The arrival of the simulator, which had two seats, prompted crews to be divided into pairs for the around-the-moon mission. With the Soyuz programme grounded, cosmonauts could be reassigned. By February 1968, five crews had been formed for the L-1 Zond mission:

- Alexei Leonov and Oleg Makarov;
- Valeri Bykovsky and Nikolai Rukhavishnikov;
- Pavel Popovich and Vitally Sevastianov;
- Valeri Voloshin and Yuri Artyukin; and
- Pytor Klimuk and Anatoli Voronov.

Although a member of the original group, Grechko had lost his place temporarily due to breaking his leg in a parachute jump (Vitally Sevastianov took his place). The definitive crews for the L-1 mission were eventually settled on 27th October 1968. The final selection was as follows:

L-1 SELECTION FOR AROUND-THE-MOON MISSION (OCTOBER 1968)

First mission: Alexei Leonov, Oleg Makarov (backup: Anatoli Kuklin).

Second mission: Valeri Bykovsky, Nikolai Rukhavishnikov (backup: Pytor Klimuk).

Third mission: Pavel Popovich, Vitaly Sevastianov (backup: Valeri Voloshin).

Not allocated: Anatoli Voronov, Yuri Artyukin, Valentin Yershov.

It is worth stressing that these selections were never absolutely final. Soviet mission assignments were frequently changed, often up till a short period before take-off, an event not unknown in the United States (e.g., Apollo 13). Nevertheless, they indicate the broad intentions which, all things being equal, would probably have happened. What was the decisive factor in the around-the-moon selection? It seems that the first two around-the-moon crews were selected for the around-the-moon flight on the basis that they would also constitute the first two landing flights. This would give them a flight to the moon and back before they went for the landing mission. This would have been like selecting Neil Armstrong and Buzz Aldrin for Apollo 11 and then deciding to send them, much earlier, on the Apollo 8 mission to the moon. Indeed, there was some discussion that the Apollo 8 crew of Borman, Lovell and Anders should, because of their around-the-moon experience, go for the moon-landing mission as well. In the event, the Americans chose, for the moon landing, men who had not flown to the moon before. The interchangeability of the Soviet around-the-moon crews with the landing crews is also reflected in the allocation to the group of mathematician scientist Valentin Yershov, one of the designers of the Zond navigation system, but whose priceless presence was also available to the moon-landing group.

MOON-LANDING TEAM

Meantime, crews were also formed for the landing mission, to fly the LOK and the LK. This included some from the L-1 group. These were also two-person crews, the commander taking the LK down to the surface, making the moonwalk and returning, while the flight engineer circled in lunar orbit. The L-1 experience in flying out to the moon and back was considered important in shaping these selections (a similar consideration was evident in American selections). For the landing mission, the first group of six was formed on 2nd September 1966:

First group of six: Yuri Gagarin, Viktor Gorbatko, Yevgeni Khrunov, Alexei Leonov, Andrian Nikolayev and Vladimir Shatalov.

Yuri Gagarin's appointment as leader was not as obvious as it looked. Following his flight in April 1961, he had spent several years as a global ambassador for the Soviet Union, a task he had performed with great aplomb. Soviet space chiefs also took the view that he was too valuable to be risked for further space missions. This was a decision he took badly and over 1963–5 he became a more problematical personality and his behaviour declined. In late 1966, he was allowed to resume training and was told he would get an early Soyuz mission, though not the first one. Eventually, he managed to win the assignment of backup to the first Soyuz mission, which eventually flew in April 1967. He took these responsibilities with his old seriousness, his health improved radically and his famous smile reappeared. This assignment meant that he would certainly command the next Soyuz mission.

At this stage there seems to have been a further dispute between those like Kamanin who wanted the cosmonauts to have a hands-on role during their mission; and chief designer Mishin who followed Korolev's view that there should be a high degree of automation. These arguments were not unknown in the American programme, though they were resolved at the earliest stage, in favour of the astronauts. By the end of 1967, no progress had been made in the provision of simulators, for they had been cancelled by Mishin. He may have considered them unimportant if most of the flight to and from the moon was under automatic control. Mishin also tried to increase the role of civilian engineers at the expense of the military. In August 1967, he now nominated a group of OKB-1 engineers for the landing mission:

OKB-1 engineer group for the landing mission: Sergei Anokhin, Gennadiy Dolgopolov, Vladimir Nikitsky (replaced by Vladimir Bugrov), Viktor Patsayev, Valeri Yazdovsky.

The first moves to form a formal moon team for the N1-L3 missions took place in October 1967. The head of the cosmonaut team, General Kamanin, had a preference for veterans and that the LK pilot should have spacewalking experience. This narrowed the field, since only one had such experience, Alexei Leonov, but more would by the time of the mission. The first landing group was selected in December 1967:

Pilots: Alexei Leonov, Andrian Nikolayev, Valeri Bykovsky, Yevgeni Khrunov, Viktor Gorbatko, Boris Volynov, Georgi Shonin, Anatoli Kuklin, Anatoli Filipchenko, Valeri Voloshin (replaced by Pytor Klimuk).

Engineers: Konstantin Feoktistov, Alexei Yeliseyev, Vladislav Volkov, Valeri Kubasov, Oleg Makarov, Vitally Sevastianov, Nikolai Rukhavishnikov, Valeri Yazdovsky, Georgi Grechko, Vladimir Bugrov.

Training began in January 1968. Some members of the group were already involved in the L-1 Zond programme. This was not seen as presenting a problem, since the landing missions were not then due until 1970–1. The same simulator problem also affected

The moon teams: Moon Team 1: Alexei Leonov, Oleg Makarov

The moon teams: Moon Team 2: Valeri Bykovsky, Nikolai Rukhavishnikov

this group. These cosmonauts did very little training because of the lack of availability of simulators.

Eventually, a 20-person lunar landing group was agreed on 13th March 1968:

Commanders: Valeri Bykovsky, Anatoli Filipchenko, Viktor Gorbatko, Yevgeni Khrunov, Anatoli Kuklin, Alexei Leonov, Andrian Nikolayev, Georgi Shonin, Valeri Voloshin, Boris Volynov.

Engineers: Konstantin Feoktistov, Georgi Grechko, Valeri Kubasov, Oleg Makarov, Vladimir Bugrov, Vitally Sevastianov, Nikolai Rukhavishnikov, Vladislav Volkov, Valeri Yazdovsky, Alexei Yeliseyev. Also assigned: Valentin Yershov.

The cosmonaut team – indeed the whole space programme – suffered a major blow in March 1968 when its leading personality died in a plane crash. Yuri Gagarin had been devastated by the death of Soyuz commander and friend Vladimir Komarov the previous year. Yuri Gagarin had been backup to Komarov and would automatically have been slated to fly the next Soyuz mission, although that had not been decided at that time. After the crash of Komarov he was grounded again, but by the end of 1967, still pressing for a flight, he had been given permission to fly again under strict conditions. On 29th March, experienced instructor Vladimir Seregin and cosmonaut Yuri Gagarin took off on a routine training flight. It seems that their MiG-15 encountered wake turbulence from the jet flow behind an unannounced MiG-21 in the area at the same time, putting their own plane into a spin. They plunged direct into the forest, killing both men outright. His funeral was the biggest there ever was in Moscow. The loss of the young, ever-popular and globally admired Gagarin was a body blow the programme could ill afford.

On 18th June 1968, the final group for the moon landing was selected:

Final group: Valeri Bykovsky, Alexei Leonov, Anatoli Voronov, Yevgeni Khrunov, Alexei Yeliseyev, Oleg Makarov, Nikolai Rukhavishnikov, Viktor Patsayev.

This was the 'landing group' for at least the following year and formed the basis of the assignment for the first two landing crews (the third is more speculative).

FINAL COSMONAUT SELECTION FOR LANDING MISSION

Prime landing crew: Leonov (commander, LK), Oleg Makarov (flight engineer, LOK).

Second landing crew: Valeri Bykovsky (LK), Nikolai Rukhavishnikov (flight engineer, LOK).

Third landing crew: Pavel Popovich (LK) and Vitaly Sevastianov (flight engineer, LOK).

Reserves: Yevgeni Khrunov (LK), Viktor Patsayev, Anatoli Voronov, Alexei Yeliseyev (LOK).

COSMONAUTS SELECTED TO LAND THE LK ON THE MOON

First landing mission: Alexei Leonov.

Second: Valery Bykovsky.

Third: Pavel Popovich.

By now, the training situation had at last improved. Leonov and others had flown adapted versions of the Mil-4 helicopter (variously called the Mil-9 and the V-10) for simulated lunar landings. These were quite hazardous, for they had to learn to land the Mils with the helicopter blades barely rotating. They would bring the Mil down to 110 m, the hover point for the lunar landing. At this stage, they would cut the engine, adjusting the blades to smoothen the touchdown. Normally, Alexei Leonov pointed out, this was a completely prohibited manoeuvre and a risky one. Bykovsky practised on a Mil-8 for several years, but hated it. A camera blister was fitted in the nose to record these landings and the film used for indoor simulations. Viktor Gorbatko alone built up over 600 hours flying helicopters. A second purpose of the helicopter landings was to test out the S-330 digital computers to be used on the LK, a test they quickly passed. A hovering vehicle, called the *Turbolets*, designed by Aram Rafaelyants, had been developed to test Soviet vertical take-off and landing planes like the Yakovlev 36. There was apparently some discussion about adapting the *Turbolets* as a lunar landing flying simulator. It may have been considered too dangerous for the cosmonauts – Neil Armstrong nearly lost his life when the temperamental American equivalent crashed – although the *Turbolets* never crashed and can still be seen in a museum today [26].

The cosmonauts went to Zagorsk to watch LK tests. Here, a full-scale model of the LK was used over a hundred times for what were called swing–drop tests: it was

Viktor Gorbatko: 600 hours of helicopter tests

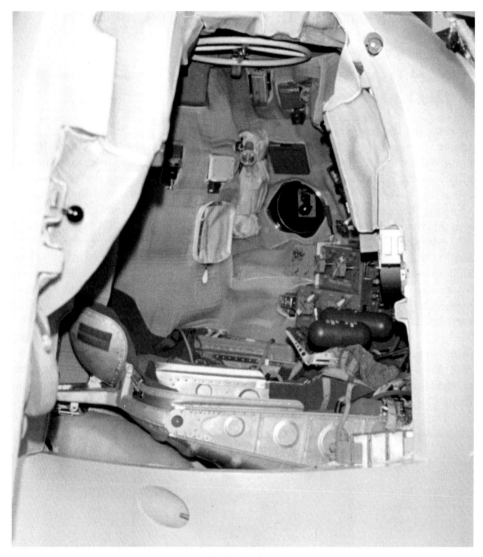

Soyuz trainer

thrown against a mockup lunar surface at various angles to see whether it would topple or not and the limits of its tolerance. Many times they went to Baikonour to watch Proton launches (though they cannot always have been heartened by what they saw). EVAs were practised in the *Kretchet* suit, both on the ground and on adapted Tupolev 104 aircraft. The moonwalk was practised on the quasi-lunar landscape around the volcanoes of the Kamchatka peninsular in the Soviet far east. A large gymnasium in Moscow's central park was converted for the practice of moonwalks, harnesses being fitted to simulate one-sixth gravity. Two expeditions of cosmonauts went into the deserts of Somalia so that they could familiarize themselves with the southern constellations that would be their main stellar point of reference during their return to Earth – mainly unflown cosmonauts were sent there so that they could not be recognized and attract American attention. They practised splashdown procedures in the Black Sea, presumably in anticipation of a water landing in the Indian Ocean (such training eventually became routine). However, the full landing simulator was not available until May 1970. Although the commanders and flight engineers trained together for many aspects of their mission, much of their training was also separate. Because only the commander would fly down to the moon and do the moonwalk, commanders did a lot of training as a group on their own. Similarly, the flight engineers had their own training programme.

ALEXEI LEONOV, THE FIRST MAN ON THE MOON

But what if Alexei Leonov had been the first man on the moon? The selection of Alexei Leonov as first man on the moon – and also to fly the first L-1 around the moon – is no surprise, for he was perhaps the leading personality of the cosmonaut squad after Yuri Gagarin himself [27].

Alexei Leonov was selected among the original group of cosmonauts in 1960. A short, well-built man full of energy and good humour, he had demonstrated his personal qualities from his teenage years. Alexei Leonov came from Listvyanka, Siberia where he was born in 1934, only months after Yuri Gagarin. He was eighth in a family of nine. Listvyanka was so cold that temperatures fell to −50°C, but the stars by night were so perfectly clear. When he was only three and in the middle of winter, his father Arkhip was declared an enemy of the people; neighbours came in and stripped their home bare and the family was evicted into the nighttime winter forest [28]. The family fled to his married sister's home. Arkhip was cleared and later rejoined them there.

Most astronauts and cosmonauts will tell you that all they ever wanted to do in life was fly a plane or a spaceship. Not Alexei Leonov, who determined to be an artist, a painter. He enrolled in the Academy of Arts in Riga in 1953. But he was unable to afford it and applied for Air Force college. He flew MiG-15s from Kremenchug Air Force Base in the Ukraine and later flew planes along the border between the two Germanies. In 1959, he was asked to go for selection for testing new types of planes and when undergoing medical tests for this unspecified assignment he met his subsequent best friend, Senior Lieutenant Yuri Gagarin.

Even before his first mission, he had two brushes with death: once when his car plunged through ice into a pond (he rescued his wife and the taxi driver from under the water) and then when his parachute straps tangled with his ejector seat (he bent the frame through brute force and freed the straps). It was a surprise to no one that he was assigned an early mission and he was Korolev's choice for the first spacewalk. To keep himself fit, in the year up to the flight he cycled 1,000 km, ran 500 km and skied 300 km. The mission, Voskhod 2, itself was full of drama. It started with triumphant success: television viewers saw him push himself away from the craft and turn head

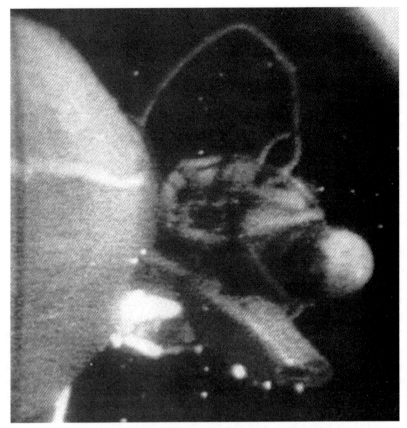

Alexei Leonov's spacewalk

over foot as he gave an excited commentary of what must have been a stunning spectacle. He had great difficulty trying to get back into his spaceship. Only by reducing the spacesuit pressure to danger level and by using his physical strength was he able to get back into his airlock. Then the retrorockets failed to fire so he and his pilot Pavel Belyayev had to light them manually on the following orbit. Instead of landing in the steppe, they came down far off course, their communication aerials burnt away, in the Urals. State radio and television played Mozart's *Requiem*, preparing the Soviet people for the worst. Their hatch jammed against a birch tree and they could barely open it. They emerged into deep snow, tapped out a morse message calling for rescuers and drew their emergency pistol to ward off prowling wolves and bears. The cosmonauts spent two nights among the fir trees while rescue crews tried to find a way of getting them out. They lit a fire to keep warm and eventually used skis to escape their ordeal. No wonder they got a hero's welcome when they returned. Definitely the Russian *right stuff*.

Alexei Leonov had an artistic bent and made many paintings of orbital flight, spacewalks, sunrises and sunsets, and spaceships landing on distant worlds. He edited

Alexei Leonov landed in a forest in the Urals

the newsletter of the cosmonaut squad, called *Neptune*, satirizing people and events with his cartoons. He maintained an extraordinary level of physical fitness and kept up his outdoor pursuits, like water skiing and hunting. He learned English and was inevitably popular with the Western media. Unlike Gagarin and Titov, he seemed to cause the commanding officers of the cosmonaut squad little trouble. Sergei Korolev praised him for his liveliness of mind, his knowledge, sociability and character. With that background and experience, he was ideally suited for the assignment of first man on the moon and, indeed, first man around the moon before that. Certainly, had he got there, the moon landing would have been well illustrated as a result. Even from his spacewalk he had generated a substantial repertoire of paintings, books and films. Unlike Neil Armstrong, who retreated for many years into academia, the extroverted Alexei Leonov would have done much to tell of his experience thereafter.

Even though Alexei Leonov did not make it to the moon, the rest of his space career was full of incident. In June 1971, he was slated to command the second mission to the Salyut space station. His flight engineer, Valeri Kubasov, was pulled only two days before the flight because of a health problem and the entire crew was replaced, despite Leonov's voluble protests. His comments became muted when the entire replacement crew was killed on returning to Earth: a depressurization valve opened in the vacuum instead of during the final stages of the return to Earth. He had cheated death again. In 1975, Leonov was the obvious choice for the joint Apollo–Soyuz mission. Leonov was the star of the show, a gracious host to the Americans on board Soyuz, cracking jokes and presenting the Americans with cartoons and souvenirs. The

Alexei Leonov splashdown-training in the Black Sea

Oleg Makarov

Americans described him as 'a really funny guy who also knows how to get us to work'. Alexei Leonov made general, was appointed commander of the cosmonaut squad from 1976 to 1982 and was a senior figure in Star Town until 1991. He still lives there, moving on to become president of one of Russia's biggest banks.

What of his companion for both missions, Oleg Makarov? Oleg Makarov was born in the village of Udomlya, in the Kalinin region near Moscow, on 6th January 1933 into an Army family. Oleg Makarov graduated as an engineer from the Moscow Baumann Higher Technical School in 1957 and worked in OKB-1 straight after. Makarov was centrally involved in the design of the control systems for Vostok, Voskhod and Soyuz, including Vostok's control panel. He was selected in the 1966 group of civilian engineers appointed to the cosmonaut squad by chief designer Vasili Mishin. Several members of this group were fast-tracked into mission assignments, and it shows that Mishin and the selectors must have thought much of him to appoint him straight away to the first moon crew with Leonov.

In the event, Oleg Makarov did get to fly into space a number of times. His first mission was to requalify the Soyuz after the disaster of June 1971. With Vasili Lazarev, he put the redesigned spaceship, Soyuz 12, through its paces in a two-day mission. The two were assigned to a space station mission in April 1975. This went badly wrong, the rocket booster tumbling out of control. They managed to separate their Soyuz 18 spaceship from the rogue rocket and after 400 sec of weightlessness made the steepest ballistic descent in the history of rocketry, the G meter jamming when they briefly reached 18 G. Their spacecraft tumbled into a nighttime valley on the border with China and they waited some time for rescue. Soyuz came down in snow, in temperatures of −7°C, the parachute line snagging on trees at the edge of a cliff. The Western media alleged that the cosmonauts had died, so on his

Valeri Bykovsky parachuting

return Makarov was sent out to play football with them to prove he was still alive. Oleg Makarov returned to space twice more. In 1978, he participated in the first ever double link-up with a space station, Salyut 6. Oleg Makarov flew again on an uneventful two-week repair mission to Salyut 6 in a new spaceship, the Soyuz T, in 1980. He died on 28th May 2003, aged 70. His obituary duly acknowledged the role he had played in the L-1 and L-3 programmes over 1965–9. Of a quieter disposition than Leonov, his technical competence must have been very evident and he would clearly have been a good selection.

What about the second crew, Valeri Bykovsky and Nikolai Rukhavishnikov? They too were slated for the second around-the-moon mission. Valeri Bykovsky was drawn from the 1960 selection with Yuri Gagarin and was given the fifth manned space mission, Vostok 5. He flew five days in orbit, three in formation with the Soviet Union's first women cosmonaut, Valentina Tereshkova. A quiet and confident man, the same age as Gagarin (born in 1934), he was a jet pilot and later a parachute instructor. He would often volunteer to test out training equipment and was the first person to try out the isolation chamber for a long period. Bykovsky left the moon group for a brief period to head up the Soyuz 2 mission, scheduled for launch on 24th April 1967, but cancelled when the first Soyuz got into difficulties. It took some time for Bykovsky to get another mission, not doing so until 1976, when he flew a solo Soyuz Earth observation mission (Soyuz 22) and then led a visiting mission to the Salyut 6 space station (Soyuz 31, 1978). After his last mission, he became director of the Centre of Soviet Science & Culture in what was then East Berlin.

Nikolai Rukhavishnikov was one of the best regarded designers of OKB-1. An intense, dedicated, serious-looking man, he came from Tomsk in western Siberia, where he was born in 1932. His parents were both railway surveyors, so he spent much of his youth on the move, living a campsite life. His secondary education was in Mongolia, and from 1951 to 1957 studied in the Moscow Institute for Physics and Engineering, specializing in transistors. Within a month of graduation, he had joined OKB-1, concentrating on automatic control systems. For the translunar mission, he

Nikolai Rukhavishnikov

planned experiments in solar physics. When the circumlunar and landing missions were delayed, he was assigned to the Salyut space station programme, being research engineer on the first mission there, Soyuz 10. Nikolai Rukhavishnikov was next selected for the Apollo–Soyuz test project, flying the dress rehearsal mission with Anatoli Filipchenko in 1974. Nikolai Rukhavishnikov was the first civilian to be given command of a Russian space mission, Soyuz 33. This went wrong, the engine failing as it approached the Salyut 6 station. Rukhavishnikov had to steer Soyuz through a hazardous ballistic descent. 'I was scared as hell', he admitted later. He later contributed to the design of the Mir space station and died in 1999.

And what about the others? The third lunar landing crew was Pavel Popovich and Vitaly Sevastianov. The two of them had worked closely on the Zond 4 mission, their voices being relayed to the spacecraft in transponder tests. Pavel Popovich came from the class of 1960, an Air Force pilot based in the Arctic. He made history in 1962 when his Vostok 4 took him into orbit close to Vostok 3 on the first group flight. An extrovert like Leonov, extremely popular, he had a fine tenor voice and sang his way through his time off in orbit. His first wife Marina was also well known, being an ace test pilot. Later, he was given command of Soyuz 14, making the first successful Soviet occupation of an orbital station, the Salyut 3. Later he became a senior trainer in the cosmonaut training centre. Vitaly Sevastianov was a graduate of the Moscow Aviation Institute and one of the teachers of the first group of cosmonauts, specializing in celestial physics. In between his own lunar training, he ran his own television programme, a science show called *Man, the Earth and the Universe*. He was one of the first of the moon group to get a mission once it became clear that there would be no early flight around the moon or landing. Vitaly Sevastianov was assigned to Soyuz 9 in 1970 and later got a space station mission, 63 days on board Salyut 4 in the summer of

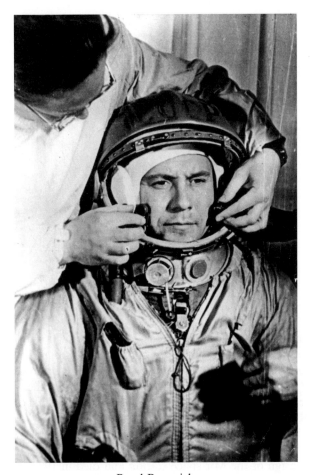

Pavel Popovich

1975, setting a Soviet record. Later, he became a leading member of the Communist Party in the Russian parliament, the Duma.

The fate of the Soviet around-the-moon and landing team makes for a number of contrasts with the American teams. For most of the American Apollo astronauts who went to the moon, the experience was the climax of their spaceflight careers and many retired from the astronaut corps soon thereafter. For the Russians, the lunar assignment was a brief period during their cosmonaut career. Although crews were named, formed and re-formed, none got close to a launch and the training experience seems to have been quite unsatisfactory. For them, the lunar assignment was short and the best of their careers was still to come. Most were quickly rotated into the manned space station programme where they went on to achieve much personal and professional success. Alexei Leonov would have made a dramatically different first man on the moon from Neil Armstrong.

Vitally Sevastianov

RETURNING TO EARTH: THE SOVIET LUNAR ISOLATION UNIT

When Neil Armstrong, Buzz Aldrin and Michael Collins returned from the moon, television viewers were amazed to see the returning heroes wrapped up in biological suits with masks and unceremoniously ushered into what looked like a camper caravan. The purpose was a serious one: to ensure that they were not contaminated with lunar soil that might in turn affect other Earthlings. The caravan was transferred to Houston where the astronauts spent the rest of their three-week quarantine.

The USSR developed a similar series of precautions and its own isolation unit. Lunar soil samples were to be received in the Vernadsky Institute of Geological and Analytical Chemistry, but a small lunar isolation unit was built in Star Town. In the Vernadsky institute, a two-floor room was set aside with two cylindrical glovebox units, each with four large viewing ports.

The opportunity to use the Soviet isolation unit came in 1970, following the mission of Soyuz 9. This was a two-man spaceflight designed to push back the then Soviet endurance record of five days and pave the way for the first Soviet space station, Salyut, due in 1970. The cosmonauts chosen, veteran Andrian Nikolayev and newcomer Vitally Sevastianov, spent 17 days in the small Soyuz cabin in June 1970.

Soon after landing, the cosmonauts were transferred to the isolation unit in Star Town by way of Vnukuvo Airport and not let out till 2 July, two weeks later: the same period of isolation as a moon journey would require. Flight debriefing was carried out behind glass partitions: telephones and microphones were used. The isolation complex had probably cost a lot to build and this was the only use it was to get. *Soviet Weekly* tried to explain:

The isolation isn't because of fears that Nikolayev and Sevastianov may have brought back strange diseases from outer space! Indeed the precautions are for the opposite reason. Doctors consider it possible that protracted space flight may lower normal immunities and they are therefore making sure that the spacemen are protected from earthbound infection until they have acclimatized.

Although Soviet spaceflights subsequently grew longer and longer, the facility was never used again. In reality, there was an element of farce about the whole episode. The *Soviet Weekly* explanation was the exact opposite of the truth, for the ultimate purpose of the unit *was* precisely to prevent infection from space-borne diseases. The real aim of the unit was never publicly revealed and we do not know what became of it subsequently. The theory behind the need for Soyuz 9 isolation had already been

Mstislav Keldysh welcomes Vitally Sevastianov home after Soyuz 9

completely undermined anyway at the point of landing. Nikolayev and Sevastianov were in weak condition when they touched down and had to be assisted from their cabin. Pictures released many years later showed them being helped and comforted, and if there had been any plans to rush them into biological protection suits, they must have been quickly abandoned. Had they indeed carried the cosmic plague with them, the entire recovery team would have been quickly infected.

REFERENCES

[1] Wotzlaw, Stefan; Käsmann, Ferdinand and Nagel, Michael: Proton – development of a Russian launch vehicle. *Journal of the British Interplanetary Society*, vol. 51, #1, January 1998.

[2] Johnson, Nicholas: *The Soviet reach for the moon – the L-1 and L-3 manned lunar programs and the story of the N-1 moon rocket.* Cosmos Books, Washington DC, 1994.

[3] Siddiqi, Asif: *The challenge to Apollo.* NASA, Washington DC, 2000.

[4] Khrushchev, Sergei: The first Earth satellite – a retrospective view from the future, in Roger Launius, John Logsdon and Robert Smith (eds): *Reconsidering Sputnik – forty years since the Soviet satellite.* Harwood Academic, Amsterdam, 2000.

[5] Lardier, Christian: Les moteurs secrets de NPO Energomach. *Air and Cosmos*, #1941, 18 juin 2004.

[6] Sanders, Berry: An analysis of the trajectory and performance of the N-1 lunar launch vehicle. *Journal of the British Interplanetary Society*, vol. 49, #7, July 1996; Sanders, Berry: An updated analysis of the three stage N-1 lunar launch vehicle. *Journal of the British Interplanetary Society*, vol. 50, #8, August 1997.

[7] Siddiqi, Asif: *The challenge to Apollo.* NASA, Washington DC, 2000.

[8] Khrushchev, Sergei: The first Earth satellite – a retrospective view from the future, in Roger Launius, John Logsdon and Robert Smith (eds): *Reconsidering Sputnik – forty years since the Soviet satellite.* Harwood Academic, Amsterdam, 2000.

[9] Pesavento, Peter and Vick, Charles P.: The moon race end game – a new assessment of Soviet crewed aspirations. *Quest*, vol. 11, #1, #2, 2004 (in two parts).

[10] Vick, Charles P.: Launch site infrastructure – CIA declassifies N-1/L-3 details. *Spaceflight*, vol. 38, #1, January 1996.

[11] Sanders, Berry: An analysis of the trajectory and performance of the N-1 lunar launch vehicle. *Journal of the British Interplanetary Society*, vol. 49, #7, July 1996; Sanders, Berry: An updated analysis of the three stage N-1 lunar launch vehicle. *Journal of the British Interplanetary Society*, vol. 50, #8, August 1997.

[12] Pesavento, Peter and Vick, Charles P.: The moon race end game – a new assessment of Soviet crewed aspirations. *Quest*, vol. 11, #1, #2, 2004 (in two parts); Slava Gerovitch: *Computing in the Soviet space programme. http://hrst.mit.edu/hrs/apollosoviet*, 2005.

[13] Ivashkin, V.V.: *On the history of space navigation development.* American Astronautical Society, history series, vol. 22, 1993.

[14] Grahn, Sven and Flagg, Richard S.: *Mission profiles of 7K-L-1 flights*, 2000. *http://www.users.wineasy.se/svengrahn/histind*

[15] Grahn, Sven and Hendrickx, Bart: *The continuing enigma of Kosmos 146 and Kosmos 154. http://www.users.wineasy.se/svengrahn/histind*

[16] Leonov, Alexei and Scott, David: *Two sides of the moon – our story of the cold war space race.* Simon & Schuster, London, 2004.

[17] Mills, Phil: Aspects of the Soyuz 7K-LOK Luniy Orbital Korabl lunar orbital spaceship. *Space Chronicle Journal of the British Interplanetary Society*, Vol. 57, Supplement 1, 2004.

[18] Pesavento, Peter and Vick, Charles P.: The moon race end game – a new assessment of Soviet crewed aspirations. *Quest*, vol. 11, #1, #2, 2004 (in two parts); additional information on the LK from Lardier, Christian: Youjnoe propose un moteur pour la lune. *Air and Cosmos*, #1987, 10 juin 2005.

[19] Pesavento, Peter and Vick, Charles P.: The moon race end game – a new assessment of Soviet crewed aspirations. *Quest*, vol. 11, #1, #2, 2004 (in two parts).

[20] Shayler, David J.: *Space suits*. Presentation to the British Interplanetary Society, 3rd June 1989.

[21] Abramov, Isaac P. and Skoog, A. Ingemaar: *Russian spacesuits*. Springer/Praxis, Chichester, UK, 2003.

[22] Abeelen, Luc van den: Soviet lunar landing programme. *Spaceflight*, vol. 36, #3, March 1994.

[23] Hendrickx, Bart: The Kamanin diaries, 1960–63. *Journal of the British Interplanetary Society*, vol. 50, #1, January 1997
  – The Kamanin diaries, 1964–6. *Journal of the British Interplanetary Society*, vol. 51, #11, November 1998.
  – The Kamanin diaries, 1967–8. *Journal of the British Interplanetary Society*, vol. 53, #11/12, November/December 2000.
  – The Kamanin diaries, 1969–71. *Journal of the British Interplanetary Society*, vol. 55, 2002 (referred to collectively as Hendrickx, 1997–2002).

[24] Da Costa, Neil: *Visit to Kaliningrad*. Paper presented to British Interplanetary Society, 5th June 1999.

[25] Marinin, Igor and Lissov, Igor: Russian scientist cosmonauts – raw deal for real science in space. *Spaceflight*, vol. 38, #11, November 1996.

[26] Gordon, Yefim and Gunston, Bill: *Soviet x-planes*. Midland Publishing, Leicester, UK, 2000.

[27] Hooper, Gordon: *The Soviet cosmonaut team* (2 vols), 2nd edition. GRH Publications, Lowestoft, UK, 1990.

[28] Leonov, Alexei and Scott, David: *Two sides of the moon – our story of the cold war space race*. Simon & Schuster, London, 2004.

6

Around the moon

The moon race between the Soviet Union and the United States climaxed in summer 1969 when the first men landed on the moon – but there was an earlier, dramatic climax six months earlier at Christmas 1968. That time the battle was to see which country would be the first to send people *around* the moon and return. Although, in retrospect, there was less and less chance that the Russians would beat the Americans to a moon landing, the chances of the Russians sending cosmonauts *around* the moon first were very real.

By late August 1968, the Russians were still trying to achieve a successful mission of the L-1 Zond around the moon. The continued troubles with the Proton rocket must have been deeply disappointing. It was then going through its most difficult phase of development and none could have imagined that it would become, much later on, one of the world's most reliable rockets. Although L-1 Zond missions had started as far back as March 1967 with Cosmos 146, none since then had been entirely successful. In August 1968, the Russians began to realize that time was no longer on their side. The first manned Apollo, redesigned after the Apollo fire, was due to make its first flight in October. Word came out of Washington that NASA was considering sending the second Apollo around the moon before the end of the year. It would be only the second Apollo flight and the first crew on the huge Saturn V rocket. The Russians had considered *four* unmanned lunar flights as essential before a manned flight: now the Americans were planning a manned flight on only the *second* Apollo mission, without any unmanned flights around the moon first.

As luck would have it, the same launch window that might take Apollo 8 to the moon opened for America on 21st December but much earlier in the USSR – from 7th to 9th December. This was entirely due to the celestial mechanics of the optimum launching and landing opportunities.

L-1 ZOND 5

Autumn was well in the air and the nighttime temperatures were cool once more when at midnight on 15th September 1968, Zond 5 rose off the pad at Baikonour and its Proton launch vehicle silhouetted the gantries, masts and assembly buildings for miles around. It all went effortlessly well, all the more remarkable after the frustrating 18 months which had passed since Cosmos 154 had triggered off so many frustrations. Sixty-seven minutes later, Zond 5 was moonbound, right on course. Its cabin contained two small turtles, fruit flies, worms and 237 fly eggs. The spacecraft weighed 5,500 kg. The plan was to recover Zond on Soviet territory after a skip trajectory, but failing that in the Indian Ocean on a ballistic return. Ten ships, equipped with three helicopters, had been sent as a recovery task force and spread out at 300 km intervals. Cameras were carried to take pictures of the close approach to the moon. Designed by Boris Rodionov of the Moscow State Institute for Geodesy and Cartography, they appear to have been developed from mapping cameras rather than from the earlier lunar missions. The standard camera for the Zond missions was a 400 mm camera taking 13 by 18 cm frames. Publicly, the official announcement said even less about the mission than usual.

Chief designer Vasili Mishin flew from Baikonour to follow the mission at the control centre in Yevpatoria. Everything was going well and the control team partied into the night. Then news came through that the stellar orientation had failed. Alexei Leonov recalled how Vasili Mishin, despite having more than enjoyed the party, analysed the problem correctly right away and had it fixed. He had good intuition, noted Leonov.

On 17th September at 6:11 a.m. Moscow time, after one failed attempt, Zond 5 successfully corrected its course at a distance from Earth of 325,000 km. At Jodrell Bank Observatory in Manchester, Sir Bernard Lovell quickly pointed his radio dish to track the enigmatic Zond 5. He picked up strong signals at once, receiving 40 min bursts on 922.76 MHz. On 19th September he was able to reveal that the spacecraft had been around the moon at a distance of about 1,950 km and was now on its way back. This information was based on the signals he had received. But nobody really knew. The Soviet Ministry of Foreign Affairs categorically denied that Zond 5 had been anywhere near the moon.

If the mission planners had been as inept as the Soviet news service, the flight would have failed at this stage. As it was, Zond 5 had seen the Earth disappear to the size of a small blue ball in the distance. Any cosmonaut then on board would have been treated to the fantastic spectacle of the moon's craters, deserts and rugged highlands sweep below him in stark profusion. Zond soared around the moon's farside and then, nearing its eastern limb, a nearly full Earth rose gently over the horizon, a welcoming beacon to guide the three-day flight home. Would a cosmonaut soon see and feel this breathtaking vista?

Early 20th September. A belated Russian admission that Zond 5 had indeed been 'in the vicinity of the moon' (as if any spacecraft happens to find itself 'in the vicinity' of the moon) was eclipsed by new, even more startling news from Jodrell Bank. A human voice had been picked up from Zond 5! Was this a secret breakthrough? Had a

Zond 5

man been aboard all along and would the Russians then announce an historic first? Not likely, said Sir Bernard Lovell. It was a tape-recorded voice, designed to test voice transmissions across deep space (one of the voices was Valeri Bykovsky's). He expected the next flight would have a cosmonaut aboard. Some 143,000 km out from Earth, Zond 5 corrected its course to adjust the entry angle. Jodrell Bank continued to track the probe till it was 80,000 km from the Earth and picking up speed rapidly. Zond 5 took its last pictures of Earth as it filled the porthole.

Zond 5 was indeed returning to the Earth. One of the reasons for Moscow's reticence was that the mission was not going well. The astro-navigation sensor had broken down, this time for good and then the gyro-platform had failed, making it impossible to restart the main engine. As a result, the two small orientation engines had to be used to set up the craft for reentry. Chances of recovery were considered slim and the gyro failure meant that a skip reentry would now be impossible. Zond 5 would now reenter steeply, ballistically.

At 6:53 p.m. Moscow time, 21st September, the L-1 cabin reached the limb of the Earth's atmosphere over springtime Antarctica, met its 10 km by 13 km reentry frame dead on, slammed into the atmosphere at 11 km/sec and burned red hot to a temperature of 13,000°C. Gravity forces built up to 16 G. After 3 min, the ordeal was over. A double sonic boom, audible over the nighttime Indian Ocean, signified survival. Still glowing, parachutes lowered the simmering Zond 5 into the Indian Ocean at 7:08 p.m. Beacons popped out to mark the location of the bobbing capsule, some 105 km distant from the nearest tracking ship. The naval vessel *Borovichy* moved in the next morning, took Zond out of the water and hoisted it aboard: in no time it was transferred to a cargo ship – the *Vasili Golovin, en route* to Bombay – where it was brought to a large Antonov air transport and flown back to the USSR. The capsule was intact, the two turtles had survived, some fly eggs had hatched and there were pictures of the Earth from deep space.

In one sweet week, all the reverses of the past 18 months had been wiped out. The moon could be *Terra Sovietica*. The first glimpse out of the porthole, the historic descriptions, the joy of rounding the corner of the moon – these could yet be Soviet successes. Zond 5 had become the first spaceship to fly to the moon and return successfully to the Earth. It was a real achievement.

All NASA could do now was cross its fingers and hope against hope that the Russians would not somehow do a manned mission first. They now knew they could. Before long the Russians released information which confirmed NASA's worst fears. They announced that Zond was identical to Soyuz, but without the orbital compartment. It had air for one man for six days. It carried an escape tower. The *Soviet Encyclopaedia of Spaceflight, 1968* rubbed it in: 'Zond flights are launched for testing and development of an automatic version of a manned lunar spaceship,' it said.

SOYUZ FLIES AGAIN

The managers of the Zond programme followed closely the requalification of the Soyuz programme, grounded since Vladimir Komarov's fatal mission in April 1967.

Zond 5 in the Indian Ocean

Zond 5's turtles

A successful flight of Soyuz would give much confidence that the closely related L-1 Zond could be flown out to the moon and back, manned, that autumn. Cosmos 238 flew a four-day profile that August to pave the way for the first manned Soyuz flight for 18 months. Selected for the mission was 47-year-old wartime combat veteran and test pilot Georgi Beregovoi.

Soyuz was ready to go on 25th October, a month after the return of Zond. So close were the American and Soviet programmes to one other at this stage that Soyuz 2 flew only three days after the end of the first test of America's new Apollo. The Americans had returned to space with Apollo 7, which orbited the Earth from the 11th to the 22nd of October, crewed by veteran Walter Schirra and novices Walter Cunningham and Don Eisele. It was technically such a perfect mission that, for the Americans, nothing now stood in the way of sending Apollo 8 around the moon. NASA's new administrator Tom Paine confirmed that Apollo 8 would fly to the moon, would make ten lunar orbits and gave the date for launch as 21st December. But he could not be certain whether his team of Frank Borman, Jim Lovell and Bill Anders would be the first to report back from there.

Soyuz 2 was launched first, on 25th October, unmanned. Soyuz 3, with Georgi Beregovoi on board, roared off the pad the next day into a misty drizzling midday sky. Half an hour later he was close to the target, Soyuz 2. Then things began to go wrong. Several docking attempts were made but, it later transpired, the craft had been aligned the wrong way up. Excessive fuel was used. Sensors failed on both spacecraft. The planned docking was abandoned, Soyuz 2 was brought down and after four days Georgi Beregovoi came home. Thick, early snow lay on the ground and the temperature was −12°C. Helicopters were in the air looking for him and villagers were outside their houses on the lookout too. Strong winds blew the capsule sideways into a snowdrift and the impact was so gentle that Georgi Beregovoi barely noticed it. Villagers waded through the snowdrifts and, amidst flecks of snow, the grinning flier had his picture taken before being whisked away for debriefing. But he had returned alive from the first successful manned orbit test of the Soyuz. Now Zond had flown automatically around the moon and a manned spaceship like Zond had circled the Earth for four days, both returning safely. A manned circumlunar mission could not be far away.

LAST TRY: ZOND 6

Just over a week later, Zond 6 headed away from Earth onto a moon trajectory (10th November). Several cosmonauts attended the launch, some hoping that one day soon they would fly a future Zond. The problems from the earlier missions then reasserted themselves. The Earth sensor failed and then the high-gain antenna failed to deploy. Despite this, two days later Zond 6 adjusted its path and on 14th November rounded the moon at a close point of 2,418 km with its automatic camera clicking away and taking metres and metres of photographs of the moon's surface. Zond 6 carried a similar payload to Zond 5: 400 mm camera, cosmic ray sensor, micrometeorite detector and some unidentified animals (probably turtles again).

Zond 6 around the moon

Even as the mission was in progress, the Russians were well aware of the start of the countdown for Apollo 8 at Cape Canaveral. The head of the cosmonaut corps, General Kamanin, considered the Apollo 8 mission to be pure adventurism and an extraordinary risk. He then considered a Russian manned flight around the moon to be a possibility for the first half of 1969. But, he conceded, the Americans might just pull off their mission first. There were hurried phone calls from the Kremlin: *Can we still beat Apollo 8?* No, said Kamanin. The best we could do is a manned flight in January, assuming Zond 6 is a success and Apollo is delayed. The most realistic date for a manned circumlunar mission was April 1969. The leaders of the space programme would not be rushed. The Soviet approach – four successful Zonds first – was clearly more conservative in respect of safety.

Zond 6 was now returning to Earth, though more problems emerged. Cabin pressure in the descent module suddenly fell, but then held. Temperatures in the fuel tanks fluctuated up and down. Zond 6 did manage to fire its engine briefly on 15th November, 251,900 km out, to adjust its course home and again a mere 10 hours before reentry to refine its trajectory. Zond 6 reentered over the Indian Ocean and dived to 45 km altitude. Pointing its heat shield at 90° to the flight path generated a cushion of lift underneath the Zond, bouncing it back into space. It was skipping across the atmosphere like a stone skipping across water, its speed now down from 11 km/sec to 7.6 km/sec. Zond 6 soared back into space in an arc and several minutes later began its second reentry. At this stage, things began to go awry. First, the high-gain antenna failed to separate. Second, more seriously, a gasket blew, depressurizing the cabin while it was still in space – which would have been fatal if unsuited cosmonauts had been on board and certainly did kill the animals on board at this stage. Third, between 3 km and 5 km above the ground, a spurious electrical signal commanded the firing of the landing retrorocket and the ejection of the parachute. Zond 6 crashed to the ground from a great height unaided. Unlike the first Soyuz, it did not explode on impact, but any spacesuited cosmonauts on board would have died. It took ground crews a day to find the cabin. They salvaged the film, which was then exhibited to the world as proof of a completely successful mission. Only decades later was film of the badly battered moon cabin released and the truth of Zond 6 told.

Zond 6 was a triumph for the skip reentry trajectory first plotted ten years earlier by Department #9. However, the performance of Zond 6 to and from the moon still needed some improvement. The landing accident was so serious that more work was still required on the landing systems, which had been considered solved by the smooth return of Soyuz 3. Sadly, the Russians did not learn their lesson as a result of the depressurization high above the atmosphere and put cosmonauts into spacesuits for such critical manoeuvres. The crew of Soyuz 11 later paid the penalty for this failure to learn. As for the planned flight around the moon by cosmonauts, it was postponed.

CLIMAX

With the flight of Zond 6, the Western press rediscovered the moon race. *Time* magazine ran a cover of an American and a Russian in a spacesuit elbowing one another out of the way as each raced moonbound. Newspapers printed cutaway drawings of 'The Zond plan' and 'The Apollo plan'. Apollo 8 astronauts Borman, Lovell and Anders were right in the middle of their pre-flight checks. Their Saturn V was already on the pad. In London, Independent Television prepared to go on air with special news features the moment Zond went up. Models and spacesuits decorated the studio. The American Navy even broke international convention to sail eavesdropping warships into the Black Sea to get close to the control centre in Yevpatoria. The Americans had Baikonour under daily surveillance by Corona spy satellites. The whole world was waiting ...

The launch window at Baikonour opened on 7th December. No manned Russian launching took place, to the evident disappointment of the Western media. They were of course unaware of the many problems that had arisen on Zond 5 or the even more serious ones on Zond 6. The Russians were in no position to fly to the moon in December 1968.

We now know that there were intense debates within the management of the Soviet L-1 programme in November and that many options were considered. The records of the time, like the Kamanin diaries, are contradictory and even confusing at

Zond over Crater Tsiolkovsky

times, but they reflected the dilemmas faced by the programme leaders. On the one hand, they did not wish to be panicked into a premature response to what they considered to be the reckless American decision to send Apollo 8 to the moon; at the same time, they realized that, with their Zond experience, they were more prepared for a lunar journey than the Americans. The cosmonauts openly expressed their willingness to make the journey to the moon ahead of Apollo 8 – several report that they sent a letter to the Politburo – but as one government minister commented, 'they would, wouldn't they.' It seems that Mishin's natural caution prevailed. Years later, Mishin told interviewers that he had recommended to the state commission against such a mission: neither Zond nor Proton were yet safe. The state commission had agreed.

The success of Apollo 8 brought mixed feelings in the Russian camp. Leaders of the Soviet space programme were full of admiration for the American achievement, which they still regarded as a huge gamble. The Americans had gone one stage further than Zond, for Apollo 8 orbited the moon ten times. The Russians were full of regrets that they had not done it first and wondered whether they had made the best use of their time and resources since the first L-1 was launched in March 1967. The head of the cosmonaut squad, General Kamanin, made no attempt to downplay the significance of the American achievement. I had hoped one day to fly from Kazakhstan to Moscow on an airplane with our cosmonauts after they had circled the moon, he confided in his diary. Now he had the sinking feeling that such a pleasurable airplane flight would never happen.

The state commission overseeing the L-1 project met on 27th December and the safe return of Apollo 8, which took place that day, was very much in their minds. They set dates for the next L-1 launches at almost monthly intervals in early 1969, starting on 20th January 1969. Some asked *What is the point?*, a moot question granted that such a mission would now achieve less than Apollo 8. The L-1 Zond programme continued, with diminishing conviction, but no one proposed cancellation.

The government's military industrial commission met three days later on 30th December. The commission underlined the value of unmanned lunar exploration and laid down a new official line on the moon programme. The Soviet Union had always planned to explore the moon by robots and would never risk lives for political propaganda. The Soviet media were invited to advertise the virtues of safer, cheaper unmanned probes. This was the first step in starting the myth, which was the official position for 20 years, that there never had been a moon race. There was little discussion of the moon-landing programme, which was now ready for the first launch of the N-1. A meeting of the party and government on the issue was set for a week later, on 8th January.

THE MISSION

What would a Russian Zond around-the-moon mission have been like? The Proton rocket would have been fuelled up about eight hours before liftoff. This is carried out automatically, pipes carrying the nitric acid and UDMH into the bottom stages, liquid oxygen and kerosene into block D. The crew – Alexei Leonov and Oleg Makarov for the first mission – would have gone aboard 2.5 hours before liftoff. Dressed in light grey coveralls and communication soft hats, standing at the bottom of the lift that would bring them up to the cabin, they would have offered some words of encouragement to the launch crews overseeing the mission. The payload goes on internal power from two hours before liftoff. The pad area is then evacuated and the tower rolled back to 200 m distant, leaving the rocket standing completely free. There may be a wisp of oxidizer blowing off the top stage, but otherwise the scene is eerily silent, for these are storable fuels. The launch command goes in at 10 sec and the fuels start to mix with the nitric acid. This is an explosive combination, so the engines start to fire at once, making a dull thud. As they do so, orange–brown smoke begins to rush out of the flame trench, the Proton sitting there amidst two powerful currents of vapour pouring out from either side. As the smoke billows out, Proton is airborne, with debris and stones from the launch area flying out in all directions. Twelve seconds into the mission, Proton rolls over in its climb to point in the right direction. A minute into the mission Proton goes through the sound barrier. Vibration is now at its greatest, as are

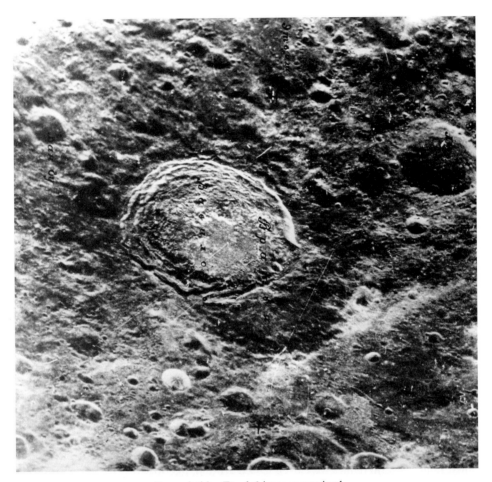

Remarkably, Zond 6 images survived

the G forces, 4 G. The second-stage engines begin to light at 120 sec, just as the first-stage engines are completing their burn. Proton is now 50 km high, the first stage falls away and there is an onion ring wisp of cloud as the new stage takes over. Proton is now lost to sight and those lucky enough to see the launch go back indoors to keep warm. Then, 334 sec into the mission, small thrusters fire the second stage downward so that the third stage can begin its work. It completes its work at 584 sec and the rocket is now in orbit.

Once in orbit, the precise angle for translunar injection is recalculated by the instrumentation system on block D. The engine of block D is fired 80 min later over the Atlantic Ocean as it passes over a Soviet tracking ship. The cosmonauts would have experienced relatively gentle G forces, but in no time would be soaring high above Earth, seeing our planet and its blues and whites in a way that could never be imagined from the relative safety of low-Earth orbit. At this stage, with Zond safely on its way to the moon, Moscow Radio and Television would have announced the

Leaving Earth, now 70,000 km distant

launching. Televised pictures would be transmitted of the two cosmonauts in the cabin and they would probably have pointed their handheld camera out of the porthole to see the round Earth diminish in the distance. The spaceship would not have been called Zond. Several names were even tossed around, like *Rossiya* (Russia), *Sovietsky Rossiya* (Soviet Russia) and *Sovietsky Soyuz* (Soviet Union), but the favourite one was the *Akademik Sergei Korolev*, dedicating the mission to the memory of the great designer.

Day 2 of the mission would be dominated by the mid-course correction. This would be done automatically, but the cosmonauts would check that the system appeared to be working properly. Although the Earth was ever more receding into the distance, the cosmonauts would see little of the moon as they approached, only the thin sliver of its western edge. Zond's dish would be pointed at Earth for most of the mission in any case.

Highlight of the mission would be at the end of day 3. Zond would fall into the gravity well of the moon, gradually picking up speed as it approached the swing-by, although this would be little evident in the cabin itself. Then, at the appointed moment, Zond would dip under the southwestern limb of the moon. At that very moment, the communications link with ground control in Yevpatoria would be lost, blocked by the moon. The spaceship would be silent now, apart from the hum of the airconditioning. For the next 45 min, the entire face of the moon's farside would fill

Earthrise for Zond 7

their portholes, passing by only 1,200 km below. The commander would keep a firm lock on the moon, while the flight engineer would take pictures of the farside peaks, jumbled highlands and craters, for the farside of the moon has few seas or *mare*. As they soared around the farside, the cosmonauts would be conscious of coming around the limb of the moon. The black of the sky would fill their view above as the moon receded below. As they rounded the moon, they would have seen a nearly full round Earth coming over the horizon, not the crescent enjoyed by Apollo 8. The *Akademik Sergei Korolev* would reestablish radio contact with Yevpatoria. This would be one of the great moments of the mission, for the cosmonauts would now describe everything that they saw below and presently behind them and as soon as possible beam down television as well as radio. Their excited comments would later be replayed time and time again.

A mid-course correction would be the main feature at the end of day 4. The atmosphere would be relaxed, after the excitement of the previous day, but in the background was the awareness that the most dangerous manoeuvre of the mission lay ahead. The course home would be checked time and time again, with a final adjustment made 90,000 km out, done by the crew if the automatic system failed. The southern hemisphere would grow and grow in Zond's window. Contact with the ground stations in Russia would be lost, though attempts would be made to retain communications through ships at sea. The two cosmonauts would soon perceive Zond to be picking up speed. Strapping themselves in their cabin, they would drop the service module and their own high-gain antenna and then they would tilt the heat-shield of their acorn-shaped cabin at the correct angle in the direction of flight. This was a manoeuvre they had practised a hundred times or more. Now they would feel the gravity forces again, for the first time in six days, as Zond burrowed into the atmosphere. After a little while, they would sense the cushion of air building under Zond and the spacecraft rising again. The G loads would lighten and weightlessness would briefly return as the cabin swung around half the world in darkness on its long, fast, skimming trajectory. Then the G forces would return as it dived in a second occasion. This time the G forces grew and grew and the cabin began to glow outside the window as it went through the flames of reentry, 'like being on the inside of a blowtorch' as Nikolai Rukhavishnikov later described reentry. Eventually, after all the bumps, there was a thump as the parachute came out, a heave upward as the canopy caught the air and a gentle, swinging descent. As the cabin reached the flat steppe of Kazakhstan, retrorockets would fire for a second underneath to cushion the landing. On some landings the cabin comes down upright, on others it would roll over. Hopefully, the helicopter ground crews would soon be on hand to pull the cosmonauts out. What a story they would have to tell! What a party in Moscow afterwards! The charred, still hot *Akademik Sergei Korolev* would be examined, inspected, checked and brought to a suitable, prominent place of reverence in a museum to be admired for all eternity.

WHAT NOW?

The success of Apollo 8 presented Soviet space planners with a double problem: how should they modify their programme in the light of America's success; and how should these changes be presented to the world? A joint government–party meeting was held on 8th January, a week into the new year. Feelings among ministers and officials verged on panic and they must now have got an inkling as to how the Americans must have felt after the early Soviet successes. Thus, a new joint resolution of the party and the Council of Ministers, #19-10, was passed on 8th January 1969. They agreed, in a bundle of decisions:

- The L-1 programme would continue, although the majority took the view that there would be little point in conducting a mission now clearly inferior to the achievement of Apollo 8.
- The programme for the N-1 would also continue, although it was apparent that it would fall short of what the Americans planned to achieve under Apollo, quite apart from running several years behind. Once successfully tested, the N-1 could be reconfigured for a mission that would overtake Apollo. Manned flights to Mars in the late 1970s were mooted – ironically the original mission for the N-1.
- Unmanned probes to the moon, Mars and Venus would be accelerated. The public presentation of the Soviet space programme would emphasize these goals.
- Ways would be explored of accelerating a manned space station programme, Vladimir Chelomei's Almaz project.

Although they now realized that their chances of beating the Americans to the moon had now sharply diminished, there was no support for the idea of abandoning the moon programme. Although this was nowhere written down, there was probably the lingering hope that America's rapid progress might hit some delays. But, in their hearts they must have known that basing their progress on the difficulties of others was not a sound basis for planning. This was not how the Soviet space programme worked in its golden years.

Now came a new generation of unmanned Russian moon probes, following the first generation (1958–60, Ye-1 to Ye-5) and the second (1963–8, Ye-6 and Ye-7). These were substantially larger and designed to be launched on the Proton rocket and called the Ye-8 series, of which the programme chief designer was Oleg Ivanovski. There were three variants:

Ye-8	Lunar rover (Lunokhod) (originally the L-2 programme)
Ye-8-5	Lunar sample return
Ye-8LS	Lunar orbiter

Although finally approved in January 1969, these missions had actually been in preparation for some time in the Lavochkin design bureau. Available first was the moon rover, or Lunokhod, the Russian word for 'moonwalker', and it was nearly ready to go. Although the Soviet Union portrayed the Lunokhod series as a cheap, safe, alternative to Apollo and although Lunokhods followed the American landings, the original purpose of the series was to *precede* and pave the way for Russian manned landings. Ideas of lunar rovers were by no means new and dated, as noticed earlier, to the 1950s. Design work had proceeded throughout the 1960s. The moon rover was intended to test the surface of the intended site for the first manned landing; later versions would carry cosmonauts across the moon. Indeed, they were endorsed in science fiction. The story of Alexander Kazanstev's *Lunnaya doroga* (Lunar road) was how a Soviet rover rescued an American in peril on the moon [1].

At the other extreme, the lunar sample return mission had been put together at astonishingly short notice. By early 1967, the design of the Ye-8 lunar rover had been more of less finished. The Lavochkin design bureau figured out that it might be possible to convert the upper age, instead of carrying a lunar rover, to carry a sample return spacecraft. The lower stage, the KT, required almost no modification and could be left as it was. Now on top sat the cylindrical instrumentation unit, the spherical return capsule atop it in turn and underneath an ascent stage. A long robot arm, not unlike a dentist's drill, swung out from the descent stage and swivelled round into a small hatch in the return cabin. The moonscooper's height was 3.96 m, the weight 1,880 kg. The plan was for a four-day coast to the moon, the upper stage lifting off from the moon for the return flight to Earth. The mission was proposed as insurance against the danger of America getting a man on the moon first. At least with the sample return mission, Russia could at least get moon samples back first. The sample return proposal, called the Ye-8-5, was rapidly approved and construction of the first spacecraft began in 1968.

Sample return missions were designed to have the simplest possible return trajectories. Originally, it was expected that a returning spacecraft would have to adjust its course as it returned to Earth. In the Institute of Applied Mathematics,

Dmitri Okhotsimsky had calculated that there was a narrow range of paths from the moon to the Earth where, if the returning vehicle achieved the precise velocity required, no course corrections would be required on the flightpath back and the cabin could return to the right place in the Soviet Union. This was called a 'passive return trajectory'. Such a trajectory was only possible from a limited number of fairly precise landing cones between 56°E and 62°E, and these were calculated following Luna 14's mapping of the lunar gravitational field. Returning from one of these cones meant that Luna could just blast off directly for Earth and there was no need for a pitch-over during the ascent, nor for a mid-course correction. If it reached a certain speed at a certain point, then it would fall into the moon–Earth gravitational field. Gravity would do the rest and the cabin would fall back to Earth. On the other hand, the passive return trajectory limited the range of possible landing spots on the moon, meant that the actual landing spot must be known with extreme precision (±10 km), the take-off must be at exactly the right second and the engine must achieve exactly the right velocity, nothing more or less [2]. Sample return missions had to be timetabled backward according to the daytime recovery zone in Kazakhstan and the need to have the returning cabin in line of sight with northern hemisphere ground tracking during its flight back to Earth. Thus, the landing time on Earth determined the landing point and place on the moon, and this in turn determined when the probe would be launched from Earth in the first place.

The Ye-8 series all used common components and a similar structure. The base was 4 m wide, consisting of four spherical fuel tanks, four cylindrical fuel tanks, nozzles, thrusters and landing legs. Atop the structure rested either a sample return capsule, a lunar rover or an instrument cabin for lunar orbit studies. By spring 1969, the time of the government and party resolution, Lavochkin had managed to build one complete rover and no fewer than five Ye-8-5s and have them ready for launch. In the case of the ascent stage, a small spherical cabin was designed, equipped with antenna, parachute, radio transmittter, battery, ablative heatshield and container for moonrock.

The first Lunokhod was prepared for launch on 23rd February 1969 and was aimed at the bay-shaped crater, Le Monnier, in the Sea of Serenity on the eastern edge of the moon [3]. The timing of the Lunokhod missions was affected by the need to land in sufficient light to re-charge the rover's batteries before the onset of lunar night. It had been arranged that when it drove down onto the lunar surface, a portable tape recorder would play the Soviet national anthem to announce its arrival. Proton failed when, 50 sec into the mission, excessive vibration tore off the shroud and the whole rocket exploded 2 sec later, the remains coming down 15 km from the launch site. For months, the military tried to find the nuclear isotope that should have powered the rover across the surface of the moon. Apparently, some local troops downrange on sentry duty found it and, clearly insufficiently briefed about the dangers of polonium radiation, used it to keep their patrol's hut warm for the rest of that exceptionally cold winter. Parts of the lunar rover were found – wheels and part of the undercarriage – and were remarkably undamaged. Even the portable tape recorder was found, playing the Soviet national anthem on the steppe, not Le Monnier bay as had been hoped [4].

Lunokhod on top of Proton

SOYUZ 4–5 REHEARSE LUNAR DOCKING, SPACEWALKING

The failure of the first Lunokhod was disappointing, for the year had otherwise started well. In mid-January, two spacecraft had been launched to Venus, Venera 5 and Venera 6, using the now-improved Molniya rocket. More important, Soyuz had flown again, rehearsing key techniques that would be used during the lunar landing mission: rendezvous, docking and spacewalking.

A rendezvous and docking of two manned Soyuz was a natural progression from Soyuz 2 and 3 the previous October and indeed the roots of the mission went back to the ill-fated Komarov flight of 1967. It was a mission absolutely essential for the manned moon landing and that was why it was in the programme. The spacewalk would simulate the transfer of the mission commander between the LOK and the LK lunar lander. However, mindful of the additional new objectives of the space programme, the mission would now be hailed as an essential step towards an orbital station instead. It was a convincing explanation for Soyuz 4 and 5 and it took in everyone at the time – except for the chiefs of NASA and one of the populist British dailies, the savvy *Daily Express*, which ran the headline 'Moon race!' the next day.

Soyuz 4 was launched first, on 14th January, with Vladimir Shatalov on board. The mission was carried out under exceptionally demanding weather conditions, in temperatures of $-22°C$ and snow around the launchpad. During mid-morning on the 15th, Vladimir Shatalov turned his Soyuz 4 towards the launch site to try and spot

Yevgeni Khrunov, Alexei Yeliseyev prepare for mission, lunar globe beside

Soyuz 5 rising to reach him. The new spaceship blasted aloft with a full complement of three men aboard: Boris Volynov, Yevgeni Khrunov and engineer Alexei Yeliseyev.

The two spacecraft approached one other during the morning of the following day. Like seagulls with wings outstretched as they escort a ship at sea, Soyuz 4 inserted its pointed probe into 5's drogue. Latches clawed at the probe, grabbed it tight, and sealed the system for manoeuvring, power and telephone. Moment of contact was 11:20 a.m. over Soviet territory. Ground controllers listened with anxiety as the two ships high above came together and met. The Soviet Union had achieved the first docking between two manned spacecraft: a manoeuvre which, it was hoped, could one day soon take place when the LK returned to the LOK in lunar orbit.

No sooner had the cosmonauts settled down after their triumph than Khrunov and Yeliseyev struggled into their *Yastreb* spacesuits. This external crew transfer was an essential feature of the moon-landing profile, being required before the descent to the moon and again on the LK commander's return. It was a slow process that could not be rushed. For his spacewalk, Leonov had already been dressed and ready to go,

Yevgeni Khrunov, Alexei Yeliseyev don spacesuits to rehearse lunar transfer

but Khrunov and Yeliseyev had to put their suits on in the orbital module cabin – as the moon-landing commander would. There was layer upon layer to put on, inner garments, outer garments, heating systems, coolant, helmets, vizors and finally an autonomous backpack. Valves were checked through, seals examined. It was not that they had not practised it enough, it just had to be right this time of all times.

Khrunov pulled a lever and the air poured out of the orbital compartment. Vladimir Shatalov had already done the same in his orbital compartment, from the safe refuge of his command cabin. The pressure gauge fell rapidly and evened off to 0. Khrunov described what happened next:

The hatch opened and a stream of sunlight burst in. The Sun was unbearably bright and scorching. Only the thick filtering vizor saved my eyes. I saw the Earth and the black sky and had the same feeling I had experienced before my first parachute jumps.

The spectacle of the two docked craft was breath-taking, he recalled. He emerged, Yeliseyev following gingerly behind, moving one hand over another on the handrails. They filmed one other, inspected the craft for damage and watched the Earth roll past below. Within half an hour they were inside Soyuz 4. They closed the hatch and repressurized the cabin. The hatch into the Soyuz 4 command cabin opened, turned like a ship's handle on a bulkhead. Vladimir Shatalov floated through and it was hugs and kisses all round. Now the Soviet Union had tested external crew transfer.

Triumph nearly turned to tragedy two days later. Soyuz 4, its crew now swollen to three, returned to Earth the following morning, coming down on hard snow in whistling winds laced with fine icy particles. The following morning was the turn of Soyuz 5 where Boris Volynov flew on, alone. First, Volynov missed his first landing opportunity due to a problem orientating the spacecraft. This was only the beginning of what could have been a very bad morning for the Soviet space programme [5]. When he did fire his retrorockets, the service module failed to separate from the descent module and Volynov's cabin instead began to go into reentry head-first, the worst possible way. Without the benefit of heatshield protection, Volynov could feel the temperature rising in his cabin. He could smell the rubber seals burning off at the top of the capsule. Knowing the end was near, he radioed details of his predicament to ground control and hastily scribbled some last notes in his log should any parts of the cabin make it to the ground. Mission control was appalled at what had happened and faced the prospect of a second Soyuz reentry fatality in less than two years. One man broke the ice a little by passing around his military hat to collect some roubles for his prospective widow, Tamara.

Awaiting his return, Tamara Volynova

Boris Volynov

Back in space, Volynov heard a sudden but welcome thump as the service module finally separated. His burning descent cabin quickly spun round and at last faced the right way, heat shield forward, for reentry. Because there had been no time to orientate the spaceship properly to use its heatshield to generate lift, he was making a steep, 9 G ballistic descent, far from the normal landing site in the southern Urals. He landed in the dark in snow, miles from anywhere, where the local temperature was −38°C. Spinning partly tangled his parachute lines and then the touchdown rockets failed to fire, so the Soyuz hit the ground with great force, breaking some of the cosmonaut's teeth. Clambering out of his still sizzling cabin, Volynov was afraid of freezing there, so he set out across the snow in his light coveralls in the direction of smoke on the horizon. The helicopter rescue crews soon found the cabin, but to their alarm, the cosmonaut was now missing! Thankfully, they were able to follow the trail of blood from his broken teeth across the snow and located him in an outhouse of local farmers. He couldn't walk for three days.

If Volynov thought his ordeal was over, he was mistaken. His next challenge was to survive a political assassination. When he and his colleagues were welcomed back to Moscow the following week in a motorcade, a young lieutenant in uniform brandishing a gun started firing at the cavalcade. He was aiming at Leonid Brezhnev, but so wildly was he firing that he got the cosmonauts' limousine instead. Its driver slumped over his wheel, dead, bleeding profusely. Beregovoi's face was splattered with blood and glass. Nikolayev and Leonov pushed Valentina Tereskhova down onto the floor to protect her. The lieutenant was grabbed by the militia and taken off to an asylum, and that was where he spent the following 20 years. The awards ceremony went ahead as planned. Putting the memory of the afternoon behind them, Russia's scientists bathed in the glow of their achievement. Mstislav Keldysh promised:

The assembly of big, constantly operating orbital stations, interplanetary flights and advances in radio, television, and other branches of the national economy lie ahead.

Mstislav Keldysh announces space stations

A few Western reporters still needled him about the moon race. There was no plan to go to the moon at the moment, he said, but when asked to confirm that Russia had abandoned plans to go the moon altogether, the ever-honest Mstislav Keldysh would not. Soyuz 4 and 5 had successfully ticked off three key elements of the Soviet lunar plan – manned docking, external crew transfer and a new spacesuit – but adroit news management portrayed the mission as part of a plan for a space station instead. As for Volynov, he took a year to recover and the doctors told him he'd never fly again. But they midjudged this brave man: he was back in training by 1972 and he did fly again.

Early Soyuz missions after the first Soyuz
27 Oct 1967   Cosmos 186
30 Oct 1967   Cosmos 188
14 Apr 1968   Cosmos 212
15 Apr 1968   Cosmos 213
28 Aug 1968   Cosmos 238
25 Oct 1968   Soyuz 2
26 Oct 1968   Soyuz 3 (Georgi Beregovoi)
14 Jan 1969   Soyuz 4 (Vladimir Shatalov)
15 Jan 1969   Soyuz 5 (Boris Volynov, Yevgeni Khrunov, Alexei Yeliseyev)

N-1 ON THE PAD

By the time of these dramatic developments in Moscow, the N-1 rocket was at last almost ready for launch. When rolled out in February, it was the largest rocket ever built by the Soviet Union, over 100 m tall and weighing 2,700 tonnes. The first stage, block A, would burn for 2 min on its 30 Kuznetsov NK-33 rocket engines. The second stage, block B, with eight Kuznetsov NK-43 engines, would burn for 130 sec and bring the N-1 to altitude. The third stage, block V, would bring the payload into a 200 km low-Earth orbit on its four Kuznetsov NK-39 engines after a long 400 sec firing. Atop this monster was the fourth stage (block G), designed to fire the lunar complex to the moon. Block G had just one Kuznetsov NK-31 engine which would burn for 480 sec for translunar injection.

For the first-ever test of the N-1, a dummy LK lunar lander had been placed on top of block D and above it, instead of the LOK lunar orbiter, a simplified version. Called the L-1S ('S' for simplified), the intention was to place the L-1S in lunar orbit and then bring it back to Earth. The L-1S was, in essence, the LOK, but without the orbital module. It still carried the 800 kg front orientation engine designed for rendezvous in lunar orbit. Calculations for the mission show that with a launch on 21st February, the spacecraft would have reached the moon on 24th February, fired out of lunar orbit on the 26th and be back on Earth by the 1st March [6]. It is intriguing that this mission would have taken place simultaneously with that of the first moonrover. Indeed, the first moonrover would have landed five hours after the L-1S blasted out of lunar orbit. In a further coincidence, 1st March was the original date the Americans had set for the launch of Apollo 9. Had the USSR pulled both these missions off, assertions about 'not being in a moon race' would have to be creatively re-explained by the ever-versatile Soviet media.

The first N-1 went down to the pad on 3rd February. It weighed in at 2,772 tonnes, the largest rocket ever built there. It was fuelled up and the commitment to launch was now irrevocable. It was a freezing night, the temperature −41°C. At 00:18 on 21st February the countdown of the N-1 reached its climax, the engines roared to life and the rocket began to move, ever so slowly, skyward. The launch workers cheered and even grisled veterans of rocket launches watched in awe as the monster took to the sky. Baikonour had seen nothing like it. Safety decreed they must stand some distance away, so they could see the rocket take off several seconds before they could hear it. Seconds into the ignition, as the engines were roaring and before it had lifted off, two engines were shut down by the KORD system, but the flight was able to continue normally, just as the system anticipated. At 5 sec, a gas pressure line broke. At 23 sec, a 2 mm diameter oxidizer pipe burst. This fed oxidizer into the burning rocket stream. This caused a fire at 55 sec which had burned through KORD's cables by 68 sec. This, shut down all the remaining engines and at 70 sec the escape system fired the L-1S capsule free, so any cosmonauts on board would have survived the failure. By then, the N-1 had reached an altitude of 27 km and, now powerless, began to fall back to Earth. The N-1 was destroyed and Alexei Leonov later recalled seeing 'a flash in the distance and a fire on the horizon'. Some of the débris fell 50 km downrange. The explosion blew windows out for miles around and Lavochkin engineers, then finishing

N-1 on the pad

preparations to send two probes to Mars, had to work from a windowless and now frozen hotel.

Despite the failure, the engineers were less discouraged than one might expect. First mission failures were not unusual in the early days of rocketry – indeed, as late in 1996, Europe's Ariane 5 was to fail very publicly and embarrassingly on its first mission. Following the report of the investigating board in March, a number of changes were made, such as taking out one of the pipes that had failed, improved ventilation and moving the cables to a place where they could not be burned. The root cause of many of the failures, though, was the high vibration associated with such a powerful rocket. This could have been identified through ground-testing, but it was too late for that now. Extraordinarily enough, American intelligence did not have satellites over Baikonour that week and completely missed the launch and the fresh crater downrange.

'LIKE STALINGRAD, BUT WITHOUT THE STUKA DIVE BOMBERS'

The next few months were difficult ones for the Soviet space programme. In March, the Russians could only watch as the Americans put the lunar module through its paces on Apollo 9. May 1969 saw the triumph of Apollo 10: Tom Stafford, Eugene Cernan and John Young had flown out to the moon, and Cernan and Stafford had brought the LM down to less than 14,400 m over the lunar surface in a dress rehearsal for the moon landing itself. Apollo 11 had been set for 16th July and the Americans had tested about all they reasonably could before actually touching down.

Summer 1969 was full of rumours of a last ditch Soviet effort to somehow upstage the American moon landing. By now, the first of the Lavochkin design bureau sample return missions of the Ye-8-5 series was ready. The first such moonscooper prepared for launch failed on 14th June 1969. The craft failed to even reach Earth orbit: an electrical failure prevented block D from firing. The Proton booster had now notched up eight failures in fourteen launches, nearly all of them mooncraft.

Time was running out for the Soviet challenge – whatever that was. In the West, observers realized there would be some challenge, though no one seemed sure exactly what. As July opened, the eyes of the world began to turn to Cape Canaveral and focused on the personalities of the three courageous Americans selected for the historic journey of Apollo 11 – Neil Armstrong, Michael Collins and Edwin Aldrin.

At this very time, Mishin's crews wheeled out the second N-1. An engineering model was also at the second N-1 pad at the time. Spectacular pictures show the two giants standing side by side just as the moon race entered its final days. Impressive though they must have been to the Russians gathered there, photographs of the two N-1s snapped by prying American spy satellites must have created near apoplexy in Washington where they panicked some American analysts to speculate on a desperate, last Russian effort to beat Apollo with a man on the moon.

As in February, the second N-1 carried another L-1S and a dummy LK. The intention was to repeat the February profile with a lunar orbit and return. Was consideration even given for a manned mission to lunar orbit to accompany the sample return mission from the surface? Assuming the same profile as February, the L-1S would have entered lunar orbit on 7th July, left for Earth on the 9th and been recovered on the 12th. Virtually all the officials concerned with the space programme converged on Baikonour for the launch. This was a heroic effort to stay in the moon race ahead of Apollo 11. One engineer later recounted that the frantic scenes reminded him of World War II in Stalingrad: 'All that was missing was the German Stuka dive bombers.'

The second N-1 lifted off very late on the night of 3rd July, at 11:18 p.m. Moscow time. Before it even left the ground, a steel diaphragm from a pulse sensor broke, entered the pump of an engine which went on fire, putting adjacent engines out of action, burning through the KORD telemetry systems and setting the scene for an explosion. KORD began to close down the affected engines: 7, 8, 19 and 20. Then an oxygen line failed, disabling engine #9. The cabling system once again disrupted, KORD shut the entire system down about 10 sec into the mission (though, for some reason, one engine continued to operate for as long as 23 sec). The N-1 began to sink back on the pad. As it did so, the top of the rocket, now 200 m above the pad, came alight at 14 sec, the escape system whooshing the L-1S cabin free just before the collapsing N-1 crashed into the base of its stand, utterly destroying the launchpad and causing devastation throughout the surrounding area. For the thousands of people watching, there was an air of surreality about it. They saw the rocket topple and fall, the fireball, the mushroom cloud but they didn't hear a thing. Then they felt the ground shake, the wind gush over them, the thunderous deafening roar and the metal rain down on top of them. Although only a few had sheltered in bunkers, none of the others had been near enough to be injured. The explosion had the force of a small nuclear explosion, toppling cars over. The physical destruction was enormous, with windows and doors blown out for miles around and little left of the pad but smouldering, gnarled girders. Part of the flame trench had even collapsed. Amazingly, the adjacent pad, with a mock N-1 rocket still installed, had survived. Even more miraculously, so had most of the crashed N-1's own tower.

The explosion was so powerful that it triggered seismographs all over the world. Days later, an American satellite flew overhead, snapping the scorch marks and devastation. When the image was received by an analyst in Washington DC he took a sharp intake of breath, stood up and yelled at the top of his voice to all his colleagues to come over and see what he had seen.

Although a preliminary investigation had guessed the cause of the disaster within a few days, the search for further clues went on for some time and the definitive report was not released for a year. The gap between this launch and the next one would inevitably be longer, as facilities must be rebuilt. Again, the failure to go for full ground-testing had proved expensive.

To Soviet space planners it was clear that the game was nearly up. Foiled by the Apollo 8 success, frustrated by one Proton and N-1 failure after another, the past two years had been marked by one misfortune after another. Nothing seemed to go right. It was a dramatic contrast to the early days when they could do no wrong and the Americans could do no right. It was the other way round now and Apollo steamed on

What the CIA saw: the N-1 pad after the explosion

from one brilliant achievement to another, dazzling the world like an acrobat who has practised a million times: except that as everyone know, NASA had not.

RUSSIA'S LAST CHALLENGE

With the failure of the N-1, Russian hopes of mounting an effective challenge to Apollo were sinking fast. The first sample return mission in June had failed and now the second N-1. Now the gambler had only one card left to play. The second Ye-8-5 was prepared and hustled to the pad in early July. The scientists may well have expected that the Proton booster would let them down again, and it was probably to their surprise that it did not. As if to scorn the earlier run of failures, it hurtled Luna 15 moonwards at 02:54 GMT on 13th July 1969. As had been the case the previous December, the celestial mechanics of the respective launch windows gave the Russians a slight advantage and enabled a launch ahead of Apollo. Once the launch was successful, preparations were put in train for a triumphant parade through Moscow, probably for the 26th or 27th July. An armoured car, covered in the Soviet flag and bedecked with flowers, would bring the rock samples from Vnukuvo Airport into Moscow, through Red Square, past the west gate of the Kremlin and on to the Vernadsky Institute where they would be displayed to a frenzy of the world's press before being brought inside for analysis [7].

The Ye-8-5

Luna 15 was the first of the third-generation Ye-8 spacecraft to succeed in leaving Earth orbit. Because it was pushing the performance of the Proton rocket to the limit, it took a fairly lengthy trajectory to the moon, in the order of 103 hours, much longer than previous moon probes. It was a tense outward journey, for telemetry indicated that the ascent stage fuel tank was overheating, threatening an explosion. Only when they turned the tank away from the sun did temperatures stabilize.

The mission profile was for a four-day coast to the moon, followed by entry into a circular 100 km lunar orbit. After a day, the orbit would be altered to bring the low point down to 16 km, right over the intended landing point. After another day, the inclination would be adjusted – probably a small manoeuvre – to ensure the lander came in over its landing site at the right angle. Sixteen hours later, after 80 hours in lunar orbit, an engine dead-stop manoeuvre would take place, after which Luna 15 would be right over the landing spot and then make a gentle final descent. After touchdown, the 90 cm long drill arm would engage. Cameras would film the scene for television. After drilling down, the arm would pop the samples back in the ascent stage. After a day on the moon, at 20:54 GMT on 21st July, Luna 15 would blast Earthward for a three-day coast to Earth. Although Luna 15 would leave the moon three hours after the American lunar module, it would fly direct back to Earth. The Americans would still face several difficult hours of rendezvous manoeuvres, transferring equipment, jettisoning the LM and then blasting out of lunar orbit, while all this time Luna 15 would speed Earthward. The Russians still faced a problem, for the return trajectory still took longer than Apollo 11 and would not get the moonrock back to Earth until 20:54 on the 24th, more than two hours after Apollo would land in the Pacific. Presumably, creative news management would have been called in to present a suitable account of the return to Earth.

Appointed to direct the mission was Georgi Tyulin. Tyulin had played an important role in the early days of the Soviet space programme. A military man, he had directed the Red Army's *Katyusha* rocket units in the war. In 1945, he was one of only four people to go to Cuxhaven, Germany, on a military delegation to watch the British fire a captured German V-2 over the North Sea, in the distinguished company of Sergei Korolev, Yuri Pobomonotsev and Valentin Glushko. He had masterminded the transfer east of the V-2 equipment to the launch base at Kapustin Yar on the Volga. Since then he had worked in military institutes, developing launch ranges and tracking systems, rising to lieutenant general.

Luna 15 produced the expected level of consternation in the West. Most observers thought Luna 15 could be a moon sample return mission, but doubted whether the USSR had the technological ability to pull it off. A typical view was this in the British *Daily Telegraph*:

While the moonshot is regarded as a last-minute attempt to detract from the American effort, it is not thought the Russians can land and bring back samples. The technical complexities are thought to be too great.

But as the Apollo 11 launching drew near – it was now only three days away – one absurd idea rivalled another. Luna 15 would jam Apollo 11's frequencies. It was there to 'spy' on Apollo 11 – like the Russian trawlers during NATO naval exercises, presumably. It was there to report back on how the Americans did it. It was a rescue craft to bring back Armstrong and Aldrin if they got stranded. With Apollo 11 already on its way to the moon, excitement about the forthcoming moon landing reached feverish levels. Scientists, experts, engineers, anyone short of a clairvoyant was called in to the television studios to comment on every change of path or signal. Cosmonaut Georgi Beregovoi, who could always be counted on to be indiscreet, let it be known that 'Luna 15 may try to take samples of lunar soil or it may try to solve the problem of a return from the moon's surface.'

By 15th July, Luna 15 was exactly halfway to the moon. Jodrell Bank – invariably tracking it – said it was on a slow course to save fuel. There was more speculation as to the ulterior motives of choosing a slow course to the moon to save fuel. Sinister implications were read into the tiniest details.

The Ye-8-5 return cabin

At 10:00 on 17th July, Luna 15 braked into lunar orbit, but entered a much wider orbit than the 100 km circular path planned, one ranging instead from 240 km to 870 km. In most subsequent official accounts of the mission, the parameters of the initial orbit were not published, although the subsequent ones were. This path was far more eccentric than what had been intended, suggesting a considerable underburn at the point of insertion into lunar orbit, one in the order of 700 m/sec rather than the 810 to 820 m/sec of all its successors [8]. There was intense radio traffic from the probe, which beamed back loud signals within 20 min of coming out from behind the moon. Jodrell Bank reported back that its signals were of an entirely new type, never heard before.

Although Moscow news sources reported that everything was normal, in fact ground control was engaged in a desperate struggle to measure the unplanned orbit and find a way to get Luna 15 into its intended path. In other circumstances, this might not have presented problems, but Apollo 11's well-publicized landing schedule was uppermost in people's minds. On 18th July, on or around the 10th orbit, ground controllers did manage to bring Luna 15 out of its highly elliptical orbit into one of 220 km by 94 km. This was still more eccentric than the 100 km orbit intended, but the perigee was close enough. The Russians had agreed to relay details of its orbit to the Americans who were worried about its proximity to Apollo 11, and they used Apollo 8 commander Frank Borman as an intermediary. Interestingly, Mstislav Keldysh told him that Luna 15 would remain in this orbit for two days (which was what had indeed been originally intended at orbital insertion), giving an orbital period of 2 hr 35 min (the one achieved after major orbital adjustment), but left it to NASA to calculate the altitude. Even today, there is a lack of a commonly agreed set of tables for Luna 15.

Manoeuvres of Luna 15
17 July	Lunar orbit insertion: 240–870 km, 2 hr 46 min, 126°
18 July	First course correction, orbit 10: 220 km by 94 km, 2 hr 35 min, 126°
19 July	Second course correction, orbit 25: 221 km by 85 km, 2 hr 3.5 min, 126°
20 July	Third course correction, orbit 39: 85 km by 16 km, 1 hr 54 min, 127°
21 July	Descent, orbit 52: 16:50 loss of signal

On 19th July, tension rose. Apollo 11, with the Apollo astronauts on board, had now slipped into lunar orbit. The world's focus shifted to the brave men on Apollo 11 carrying out their final checks before descending to the surface of the moon. Now on its 39th orbit, Luna 15 fired its motor behind the moon to achieve the pre-landing perigee of 16 km. This was its final orbit, for at 16 km there was barely clearance over the mountain tops and was about as low as an orbit could go. The probe could only be preparing to land. The perilune was known to be over the eastern edge of the moon, not far from the Apollo landing site in the Sea of Tranquillity, but farther to the northeast, over a remarkably circular *mare* called the Sea of Crises. The Luna 15 mission was back on course.

Luna 15 and Apollo 11: timelines

	Luna 15	Apollo 11
13	02:54 Launch	
16		13:32 Launch
17	10:00 Lunar orbit insertion	
18	13:00 Apolune lowered to 220 km	
19	13:08 Perilune to 85 km	17:22 Lunar orbit insertion
20	14:16 Final orbit, perilune 16 km	
	[19:00 Original scheduled landing]	20:19 Landing on moon
21	15:50 Loss of signal on landing	
	[20:54 Original scheduled lunar liftoff]	17:54 Take-off from the moon
22		04:57 Leave lunar orbit
23		
24	[20:54 Original scheduled landing]	16:50 Splashdown

Note: times are GMT.

In reality, Luna 15 was now in fresh trouble. When the engineers turned the radar on at the low point of the orbit, 16 km, to verify the landing site, they got problematic readings. Although the Sea of Crises has a flat topography – some of the moon's flattest – the radar instead indicated quite an uneven surface. Luna 15 was scheduled to land at 19:00 that evening, the 20th, only an hour before Apollo 11's *Eagle*, coming into the Sea of Crises from the north. Tyulin decided to delay the landing for 18 hours in order to retest the radar, try and get a clearer picture of the terrain and calculate the precise moment for retrofire as carefully as possible. This must have been a difficult decision for, by doing so, there was no way that Luna 15 could be back on Earth before Apollo 11. This was the first time that Russia had attempted a soft landing *from lunar orbit* (indeed, the same could be said for Apollo 11's *Eagle*). The retrofire point had to be precisely set in altitude and location: 16 km above the surface, not more than 19 km, not less than 13 km, so as to match the capacity of the engine.

Few people gave much thought to Luna 15 for the next day as they listened in wonder to the descent of Neil Armstrong and Edwin Aldrin to the lunar surface, agonized through the final stages of the descent and then watched the ghostly television images of the two men exploring the lunar surface. On the early evening of 21st July, Armstrong and Aldrin stood in their lunar module going through the final checks before take-off from the moon, a manoeuvre that had never been done before. Just as they did so came a final newsflash from Jodrell Bank. It was to serve as Luna 15's epitaph:

Signals ceased at 4.50 pm this evening. They have not yet returned. The retrorockets were fired at 4.46 pm on the 52nd orbit and after burning for four minutes the craft was on or near the lunar surface. The approach velocity was 480 km/hr and it is unlikely if anything could have survived.

Jodrell Bank identified the Sea of Crises as the landing spot. The dramatic conclusion to Luna 15, just as the lunar module was about to take off, made for great television drama. Imagine, though, if Luna 15 had been able to follow its original schedule, land just before *Eagle* and take off just afterwards: this was a script beyond the imagination of Hollywood.

Despite his caution and giving the landing his best shot, Tyulin's Luna 15 impacted 4 min into a 6 min burn when it should have still been 3,000 m above the surface. Official explanations ventured that it hit the side of a mountain. Granted that the Sea of Crises is one of the flattest *maria* on the moon, this seems implausible. More likely, there was a mismatch between the low point of the orbit, 16 km and the imagined surface point (a surface reference point can be difficult to calculate when there is no natural marker, like sea level on Earth). A navigation error was most likely responsible. Another explanation is that the landing motor was late in firing [9]. American military trackers kept a close watch on Luna 15, and their analysis indicated that the Russians had difficulty controlling the pitch axis on Luna 15. Thirty-five years later, their reports strangely remained 'top secret'.

Many, mostly unconvincing reasons were advanced by the Soviet press to explain away Luna 15. One publication even had the nerve to claim that 'if it hadn't happened to coincide with the dramatic Apollo lunar flight, it would hardly have received a mention at all.' So what was Luna 15 then? Just a new moon probe. A survey ship that was highly manoeuvrable. Indeed, it had a flexibility that the American moonship did not have because it could manoeuvre freely, unlike Apollo which was stuck in narrow equatorial orbit. One wonders if the author – one 'Pyotr Petrov' – even believed this himself.

Following the first moon landing, the original Apollo lunar exploration programme was cut back and redirected. The Russian programme, for its part, went through a prolonged and painful reorientation before eventual cancellation. The programme of unmanned lunar exploration was the only substantial part salvaged from its protracted demise. The redirection of the Soviet moon programme may be divided into several phases:

- Winding down of the L-1 Zond around-the-moon programme, 1969–70.
- Testing the LK and the LOK, 1971–2.
- Cancellation of the original N-1 moon-landing programme in 1971.
- Replacement by a revised scheme of lunar exploration, 1971–4, the N1-L3M.
- Suspension of the N-1 in 1974, with its final cancellation in 1976.

WINDING DOWN THE L-1 ZOND AROUND-THE-MOON PROGRAMME

Even though Apollo 8 had flown around the moon in December 1968, the L-1 programme was not abandoned. There were several reasons. The hardware had been built or was still in construction. So much investment had gone into the programme that it was felt better to test out the technical concepts involved than write them off altogether and deny oneself the benefits of the design work. If these tests went well, a manned moon circumlunar mission could still be kept open as an option. Indeed, with some of the political pressure lifted, designers looked forward to testing their equipment without the enforced haste required by American deadlines. There was also an official problem, bizarre to outsiders, which was that the Soviet government lacked a mechanism to *stop* the moon programme. At governmental level, no one was yet prepared to admit failure or to take responsibility for what had gone wrong [10]. The resolutions of August 1964 and February 1967 remained in effect, unrepealed. According to Alexei Leonov, the government decided that if the next Zond succeeded, then the following one would be a manned flight, even after Apollo 8.

A new Zond was readied in January and left the pad on 20th January 1969. It is unclear what profile it would have flown, for it was outside the normal launch window. The cabin used was the one salvaged from the April 1968 failure [11]. The second stage shut down 25 sec early at 313 sec, but the other second-stage engines completed the burn. During third-stage firing, the fuel pipeline broke down and the main engine switched off at 500 sec, triggering a full abort. The emergency system lifted the Zond cabin to safety, and it was later retrieved from a deep valley near the Mongolian border. As we know, the period January to July 1969 lacked good launch-and-return windows for Zond missions around the moon, so any missions would have to be performed either under less than ideal tracking, transit or lighting conditions, or would have to be fired at a simulated moon, which was probably the case this time.

There were still Zond spacecraft available. At this stage, a perfect circumlunar flight was still required before a manned mission could be contemplated. However, a Russian manned circumlunar flight would now, after Apollo 11, make an even more

A full Earth for Zond 7

invidious comparison after Apollo 8. The chances that the cosmonauts would be allowed to fly were fading.

The Russians took advantage of the first of the new series of lunar opportunities opening in the autumn. Zond 7 left Baikonour on 8th August 1969, only two weeks after Luna 15's demise and at about the time that the Apollo 11 astronauts were emerging from their biological isolation after their moon flight. Thirty turtles had been ready for the mission and four were selected. Zond 7 was the only one of the series to carry colour cameras. Cameras whirred as Zond skimmed past the Ocean of Storms and swung round the western lunar farside 2,000 km over the Leibnitz Mountains. Zond 7 carried a different camera from its predecessors, a 300 mm camera with colour film taking 5.6 cm^2 images. Strikingly beautiful colour pictures were taken of the Earth's full globe over the moon's surface as Zond came around the back of the moon. Like Zond 5, voice transmissions were sent on the way back. Zond 7 headed back to the Earth, skipped like a pebble across the atmosphere to soft land in the summer fields of Kustanai in Kazakhstan after 138 hr 25 min. It was a textbook mission.

How easy it all seemed now. After the total success of Zond 7, plans for a manned circumlunar mission were revived and there were still four more Zond spacecraft in the construction shop – one even turned up in subsequent pictures with 'Zond 9' painted in red on the side. The state commission responsible for the L-1 Zond programme met on 19th September and the decision was taken to fly Zond 8 as a final rehearsal around the moon in December 1969, with a manned mission to mark the centenary of Lenin's birth in April 1970, which would be a big national event.

This plan, which was probably designed to appeal to the political leadership, did not in fact win government approval. There were mixed opinions among those administering the Soviet space programme as to whether a man-around-the-moon programme should still fly. Many had serious reservations about flying a mission that would be visibly far inferior not only to Apollo 11 but to the two Apollo lunar-orbiting flights that preceded it. Others disagreed, arguing that the Soviet Union would, by sending cosmonauts to the moon and back, demonstrate at least some form of parity with the United States. In 1970, few other manned spaceflights were in prospect, so a flight around the moon would at least boost morale. The normally cautious chief designer Vasili Mishin pressed hard for cosmonauts to make the lunar journey on the basis that the experience gained would be important in paving the way for a manned journey to a landing later. The political decision, though, was a final 'no', the compromise being that Mishin was allowed to fly one more Zond but without a crew. Two of the cosmonauts in the programme subsequently went on record to explain the decision. The political bosses were afraid of the risk that someone would be killed, said Oleg Makarov, who was slated for the mission. Another cosmonaut involved, Georgi Grechko, felt that the primary reason was political: there was no point in doing something the Americans had already done [12]. In the end, Lenin's centenary was marked, indirectly and two months after the event, by the 18-day duration mission of Andrian Nikolayev and Vitally Sevastianov.

Zond 8 was eventually flown (20th–27th October 1970). It carried tortoises, flies, onions, wheat, barley and microbes and was the subject of new navigation tests. Astronomical telescopes photographed Zond as far as 300,000 km out from Earth to check its trajectory. Zond 8 came as close as 1,110 km over the northern hemisphere of the lunar surface, the closest of all the Zonds. Two sets of black-and-white images were taken, before and after approach. The 400 mm black-and-white camera of the type used on Zonds 5 and 6 was carried. These were high-density pictures, 8,000 by 6,000 pixels and are still some of the best close-up pictures of the moon ever taken [13].

There have been contradictory views as to whether Zond 8 was intended to return to the Soviet Union or be recovered in the Indian Ocean. The records now show that the recovery in the Indian Ocean was deliberate and not the result of a failure. As we know, the optimum trajectory for a returning Zond was to reenter over the southern hemisphere and make a skip reentry, coming down in the normal land recovery zone (Zond 6 and 7), or, if the skip failed, a ballistic descent into the Indian Ocean (Zond 5).

The alternative approach, one favoured by Mishin, was to come through reentry over the northern hemisphere, with good contact with the ground during this crucial period, but make a southern hemisphere splashdown. This route had not been tried before. Two Soviet writers of the period confirm that the purpose of Zond 8 was indeed 'to make it possible to verify another landing version with deceleration over the USSR' [14]. Zond 8 made a smooth northern hemisphere skip reentry and came down in the Indian Ocean 24 km from its pinpoint target where it was found within 15 min by the ship *Taman*. This seemed to prove Mishin's point. Six years later, though, cosmonauts Vyacheslav Zudov and Valeri Rozhdestvensky splashed down in a lake and very nearly drowned during a protracted and hazardous recovery.

Analysis of the biological samples found similar results across the series. The turtles were hungry and thirsty after their return: hardly a surprise as they had not been fed or watered during their mission. They were examined for changes to their heart, vital organs and blood. There were some mutations in the seeds as a result of radiation. Overall, radiation dosages seemed to be well within acceptable limits, not posting a danger to cosmonauts and not significantly different from conditions in Earth orbit.

Thus, of nine Zond missions and of six attempts to fly to the moon, only Zond 7 and 8 were wholly successful. The last two production Zonds were never used. Just as the Russians tested their lunar hardware in Earth orbit successfully (Cosmos 379, 382, 398, 434), they tested their round-the-moon hardware successfully. We now know that the Russians reached the stage where they *could*, with a reasonable prospect of success, have proceeded to a manned around-the-moon flight. Years later, Vasili Mishin was asked about his period as chief designer and whether he would have done things differently. 'Perhaps,' he said wistfully, 'I would have insisted on making a loop around the moon, even after the United States, because we had everything ready for it. Maybe we could have done it even before the Americans' [15].

L-1, Zond series

10 Mar 1967	Cosmos 146
8 Apr 1967	Cosmos 154 (failure)
28 Sep 1967	Launch failure
23 Nov 1967	Launch failure
2 Mar 1968	Zond 4
23 Apr 1968	Launch failure
22 Jul 1968	Pad accident
15 Sep 1968	Zond 5
14 Nov 1968	Zond 6
20 Jan 1969	Launch failure
8 Aug 1969	Zond 7
20 Oct 1970	Zond 8

L-1/Zond series: scientific outcomes
- Characterization of Earth–moon, moon–Earth space.
- Mapping of lunar farside.
- Acceptability of radiation limits for biological specimens.

TESTING THE LK AND THE LOK

Other Soviet equipment for the moon landing was tested. Would the Russian lunar module have worked? Yes, it probably would have, for in 1970–1 the LK was put through a series of exhaustive tests in Earth orbit which it passed with flying colours. Block E of the lunar module had been tested in Zagorsk 26 times, but never in flying conditions.

These were called the T2K tests. The lunar lander, the LK, was tested without its landing legs, since these were primarily propulsion tests of the block E system with its 2.05-tonne thrust, intended to simulate the two major burns of the lunar surface landing and then the subsequent ascent to orbit. The Russians did three tests, all unmanned – while, many years earlier, the Americans had also carried out three (Apollo 5, January 1968, unmanned; Apollo 9, March 1969 and Apollo 10, May 1969, both manned).

The first T2K was launched by a Soyuz rocket on a sunny morning, 24th November 1970, under the designation Cosmos 379, witnessed by its designers. It entered orbit of 192 km to 230 km. On 27th November, after simulating the three-day journey to the moon, the LK fired its variable throttle motor to simulate the lunar landing, descent and hovering over the moon's surface (250 to 270 m/sec ΔV), changing its orbit to an apogee of 1,120 km. On the 28th, after simulating a day on the surface of the moon, as it were, the LK fired its engine again to model the lunar ascent. Everything went perfectly. This was necessarily a powerful burn, 1,320 to 1,520 m/sec ΔV). Cosmos 379 ended up in a 14,300-km high orbit, eventually burning up in September 1983.

Further tests of the LK moon cabin were made by Cosmos 398 (26th February 1971) and Cosmos 434 (12th August 1971). On each mission, the landing frame was left in an orbit of 120 km, the ascent cabin much farther out. Cosmos 398 crashed into the South Atlantic in December 1995. In the case of Cosmos 434, the final orbit was 186 by 11,834 km. Unlike the American lander, the landing frame had no propulsive engine in its own right.

The end of the Cosmos 434 mission had a treble irony. Only days after its conclusion, the N1-L3 plan for landing on the moon was cancelled as Mishin persuaded the government to go for a more ambitious lunar-landing plan using a different method, the N1-L3M. Second, that October LK designer Mikhail Yangel invited guests to attend his 60th birthday party, but he died suddenly just as they began to arrive at his home. Hopefully, he realized before his death just what a fine lunar module he had designed and built. Third, in August 1981, Cosmos 434 began to spiral down to Earth. Only three years earlier, a nuclear-powered surveillance satellite had caused a scare when it began to tumble out of orbit. This time, the Soviet Union assured the world there was no need to worry since, because Cosmos 434 was 'a prototype lunar cabin', it had no nuclear fuel. This was the first time the Soviet Union had ever publicly admitted, although inadvertently, to the existence of its manned moon-landing programme.

Thankfully, these orbiting Cosmos were not the only LKs completed. Examples of the LK can still be found: in the Moscow Aviation Institute; the Mozhaisky Military Institute of St Petersburg; and at the home of its builder, now called NPO Yuzhnoye, in Dnepropetrovsk. NPO Yuzhnoye has an exhibit of its great engine. And for those contemplating a return to the moon, Yuzhnoye has kept the blueprints too.

The LK tests

24 Nov 1970	Cosmos 379
26 Feb 1971	Cosmos 398
12 Aug 1971	Cosmos 434

The LK manoeuvres

Cosmos 379	24 Nov 1970	51.61°	191–237 km	
1st manoeuvre		51.63°	192–233 km	
		51.65°	296–1,206 km	263 m/s
2nd manoeuvre		51.59°	188–1,198 km	
		51.72°	177–14,041 km	1,518 m/s
Cosmos 398	26 Feb 1971	51.61°	191–258 km	
1st manoeuvre		51.61°	189–252 km	
		51.6°	186–1,189 km	252 m/s
2nd manoeuvre		51.6°	186–1,189 km	
		51.59°	200–10,905 km	1,320 m/s
Cosmos 434	12 Aug 1971	51.6°	189–267 km	
1st manoeuvre		51.6°	188–267 km	
		51.6°	190–1,261 km	266 m/s
2nd manoeuvre		51.6°	188–1,262 km	
		51.54°	180–11,834 km	1,365 m/s

Source: Clark (1988)

What about the LOK and block D? Granted that a working version of the LOK was never successfully launched, it is impossible to comment on its performance. With the flight of Zond around the moon (1969–70) and the requalification of Soyuz (1968–9), it is reasonable to presume that it would have been a successful spacecraft. The first LOK was scheduled to be tested on the fourth flight of the N-1 in 1972. In the meantime, it was decided to proceed with tests of block D for its lunar orbit mission. The types of manoeuvres planned for block D had, unlike Zond and Soyuz, not been tested. Block D engine firings were required for mid-course corrections outbound, to put the complex in lunar orbit and then, second, carry out the powered descent initiation down to 1,500 m over the surface. They were carried out with a block D attached to a modified Zond and called the KL-1E ('E' for experimental).

The first, on 28th November 1969, failed when the first stage of the Proton exploded. The second was Cosmos 382, sent aloft on 2nd December 1970. The manoeuvres simulated the lunar orbit insertion burn, course corrections and the powered descent, respectively. All seem to have gone perfectly. Cosmos 382 aroused some interest at the time. Western experts could not understand why the Russians were flying spacecraft in lunar-type manoeuvres long after Russia had lost a moon race it now claimed it had never been part of.

Block D tests, 1969–70

| 18 Nov 1969 | KL-1E test: failure |
| 2 Dec 1970 | KL-1E test: Cosmos 382 |

Manoeuvres of Cosmos 382 L-1E

Launch	2 Dec 1970	51.6°	190–300 km	
1st manoeuvre	3 Dec 1970	51.6°	190–300 km	
		51.57°	303–5,038 km	986 m/s
2nd manoeuvre	4 Dec 1970	51.57°	318–5,040 km	
		51.55°	1,616–5,071 km	288m/s
3rd manoeuvre	8 Dec 1970	51.55°	1,616–5,071 km	
		55.87°	2577–5,081 km	1,311 m/s

Source: Clark (1988, 1993)

Following the success of Soyuz 4 and 5, a further manned Earth orbital test of the lunar orbit rendezvous manoeuvre was planned, similar to those which the United States carried out on Apollo 9. This was called the *Kontakt* mission, and its specific purpose was to test the rendezvous mechanisms of the LOK and the LK lunar lander. *Kontakt* was the docking system that would have been used had the original Soyuz complex gone ahead. *Kontakt* was developed by Alexei Bogomolov of the Moscow Engineering Institute. It might earlier have been used for Soyuz Earth orbital missions, but a rival system called *Igla* was adopted instead. *Kontakt* came back into the reckoning for the manned lunar landing, being adopted for the programme partly on account of its simplicity.

Tests of the *Kontakt* system in Earth orbit were clearly essential before it was committed to lunar orbit rendezvous. These were planned for 1970 and two Soyuz were readied for the mission, one active, one passive. The active crew was Anatoli Filipchenko and Georgi Grechko, the crew for the passive Soyuz was Vasili Lazarev and Oleg Makarov. The mission was assigned high priority, with up to 16 cosmonauts being put through the training programme for the mission. A second, follow-up double mission seems also to have been envisaged. Bogomolov's delivery of the *Kontakt* system, originally for 1970, kept slipping. In August 1971, the LOK and LK were abandoned and the missions were formally terminated in October 1971. The four Soyuz in preparation were dismantled and the parts used for other missions.

CANCELLATION OF THE MOON-LANDING PROGRAMME

Zond 8 marked the end of the around-the-moon programme. The landing programme, dependent on the testing of the N-1, still continued. Chief Designer Vasili Mishin continued to enjoy support at the highest level in the Politburo, especially from Andrei Kirilenko. Testing of the N-1 continued with a view to its completing its original purpose, or, alternatively, to carry large payloads to low-Earth orbit. After the two disasters of 1969, KORD was redesigned. The system could no longer be closed down entirely during the first 50 sec of flight. A fire-extinguishing system, using freon gas, was installed. The NK-33 engines were tested more rigorously, with new systems for quality control. Filters were installed to stop foreign objects from getting into the engines. Cabling was better protected against fire. Pumps were improved. The launchpads were rebuilt.

The third N-1 was ready to fly again two years later. Unlike its two predecessors, it was not aimed at the moon, carrying only a dummy LK and LOK (and a dummy escape tower). Launching took place at night on 27 June 1971, while, incidentally, three cosmonauts were aloft in the Salyut space station (at one stage, it had been planned for them to look out for the launch). From as early as 7.5 sec, the vehicle began to roll about its axis. By 40 sec, the small vernier engines lost the ability to counteract the roll and at 45 sec the rocket began to break up, the payload falling off first. At 51 sec, the redesigned KORD system shut the lower rocket stage down. The stages separated and the rocket crashed to destruction, the first stage gouging out a 30 m crater 20 km downrange. The escape tower was a mockup and did not fire. Ironically, the failure of the rocket's roll control system was due to the fact that all the engines of the N-1 were actually firing together at take-off at the same time for the first time, none having been shut down by KORD. The thrust of the 30 engines, all firing together, created a strong roll effect that the vernier engines had been insufficiently strong to counter. Had all the engines fired properly at launch on the first or second take-offs, this would have been apparent then. Or, more to the point, the problem might have been identified if there had been proper ground-testing.

The failure of the third N-1 took place at a time when the Americans were making rapid progress in lunar exploration. In February 1971, the Americans had returned to the moon with Apollo 14 and were about to proceed to the three-day surface missions of Apollos 15, 16 and 17. In July 1971, the Americans landed at Mount Hadley and spent three days there, driving a rover around the mountains and to the edge of a rille. The old N1-L3 plan, putting only one cosmonaut on the lunar surface for only a few hours, looked ever more inadequate. The old N1-L3 plan was finally terminated in August 1971, at the time the third LK was successfully being put through its paces as Cosmos 434. The termination of the plan permitted remaining LK and LOK hardware to be flown, presumably on N-1 testflights.

REDIRECTION: THE N1-L3M PLAN

The Soviet plan for lunar exploration was now decisively redirected. Vasili Mishin now devised a moon plan even more ambitious than that of Apollo. He decided to match the *three days* of *two* Apollo astronauts on the moon with a Soviet plan to put *three* cosmonauts there *for a month*. The new Mishin plan, called the L-3M ('M' for modified) envisaged a manned lunar mission with two N-1 rockets. The N-1 would be upgraded with a more powerful hydrogen-powered upper stage. The exact date on which the L-3M plan was adopted is uncertain. The programme was first mooted in September 1969, clearly a first response to the American moon landing two months earlier, and the title 'L-3M' first appeared in print in documents in January 1970. The project was scrutinized by an expert commission under Mstislav Keldysh in spring 1971, and a resolution of the chief designers *Technical proposals for the creation of the N1-L3M complex* was signed off on 15th May 1972.

The first N-1 would place a large 24-tonne lunar lander descent stage, the GB-1, based on block D, in lunar orbit. Independently, a second N-1 would deliver a three-man lunar lander and return spacecraft, GB-2, to link up with the descent stage. Together they would descend to the lunar surface. Initially, three cosmonauts would work on the moon for a full lunar day (14 Earth days) but this would be later extended to be a month or longer. Eventually, four cosmonauts would live on the moon for a year at a time. The ascent stage would have a mass of 19.5 tonnes on launch from the moon and 8.4 tonnes during trans-Earth coast. Launch would be direct back to Earth, like Luna 16, without any manoeuvres in lunar orbit. The lander would incorporate Soyuz within what was called a cocooned habitation block, or OB, a sort of hangar. The crew could climb out of Soyuz into the hangar, put on their spacesuits there and use the hangar as a pressurization chamber before their descent to the lunar surface. The Americans might be first to the moon, but the Soviet Union would build the first moon base. Mishin envisaged the dual N-1 mission taking place in the late 1970s. Mishin's new plan even won the approval of long-time N-1 opponent, Valentin Glushko. At one stage, the Soviet military considered turning the moon base into their first military headquarters off the planet [16].

An important feature of the N1-L3M was the redesign of the N-1 launcher, given the tentative name of the N1-F (industry code 11A52F). The airframe was much improved and there was a hydrogen-powered upper stage. The top part of the rocket, needle-shaped for the early N-1, was now bulkier and broader. The fact that Russia successfully developed a hydrogen-powered upper stage during the 1960s was one of the last, well-kept secrets of the moon race. The West had not believed the Russians capable of such a development, and it did not come to light until India bought a hydrogen-powered upper stage from the Russians in the 1990s. In fact, we now know that Russia had worked on hydrogen propulsion from 1960 onward and that hydrogen-powered stages had been part of the 1964 revision of the Soyuz complex in OKB-1. This research had continued to progress and by the late 1960s was reaching maturity. Linking this research to the new, improved N-1 made a lot of sense.

The hydrogen motor was the KVD-1, built by the Isayev design bureau (KVD stands for *Kislorodno Vodorodni Dvigatel*, or oxygen hydrogen engine). The role of the KVD-1 was to brake the assembly into lunar orbit and make the descent to the lunar surface. The KVD-1 engine had a burn time of 800 sec and a combustion chamber pressure of 54.6 atmospheres. The KVD-1 had a turbopump-operated engine with a single fixed-thrust chamber, two gimballed thrust engines, an operating period of up

Alexei Isayev

to 7.5 hours and a five times restart capability. It weighed 3.4 tonnes empty and 19 tonnes fuelled. Its thrust was 7,300 kg and the specific impulse was 461 sec, still the highest in the world at the end of the century. It was 2.146 m tall, 1.28 m diameter and weighed 292 kg. It was sometimes called block R and had the industry code of 11D56.

The Isayev bureau was one of the least well-known of all the Soviet design bureaux and featured little in the early *glasnost* revelations about the Soviet space programme, its design bureaux and rocket engines. The bureau started life as Plant #293 in Podlipki in 1943, directed by one of the early Soviet rocket engineers, Alexei Isayev. Born 11th October 1908 (os) in St Petersburg, he was a mining engineer and had been given his own design bureau in 1944. This was renamed OKB-2 in 1952, being given its current name, KM KhimMach, in 1974. Besides spacecraft, its work has concentrated on long-range naval, cruise and surface ballistic missiles and nuclear rockets, and by the early 1990s had built over a hundred rocket engines, mainly small ones for upper stages, mid-course corrections and attitude control.

The KVD-1 prototype was first fired in June 1967. The engine was later tested for 24,000 sec in six starts. Five block R stages were built and tested over the years 1974–6 and the engine was declared fully operational. In fact, the KVD-1 was not the only Soviet hydrogen-powered upper stage. Nikolai Kuznetsov also struggled with a hydrogen-powered upper stage engine called the NK-15V, with a thrust of 200 tonnes, which would replace block B. OKB-165 of Arkhip Lyulka also developed engines for the third stage and fifth stage, respectively, 11D54 and 11D57 or block S. A scale model was built of a revised N-1 with hydrogen upper stages [17]. Approval was given for these developments in June 1970.

A new engine and new fuel were developed for the N1-L3M lunar module. Here, under Vasili Mishin, Valentin Glushko's OKB-486 design bureau made a belated appearance in the N-1 programme. Valentin Glushko designed the new RD-510 engine, with 12 tonnes thrust [18]. The fuel was hydrogen peroxide, also called High

Test Peroxide (HTP). Only one other country in the world used hydrogen peroxide for its space programme: Britain, for its Black Arrow rocket. Hydrogen peroxide actually went back to wartime Germany where it had been developed by Dr Hellmuth Walter for high-speed U-boats.

Like Glushko's favourite fuel, nitric acid, hydrogen peroxide could be kept at room temperatures for long periods. Hydrogen peroxide had one advantage over nitrogen-based fuels: it did not require the mixing of a fuel with a oxidizer. It was a monopropellant, requiring one tank and a means of igniting the rocket (metallic filings were inserted). There was no need to mix in the product of two tanks in a very precise ratio to get the desired thrust. Nor was HTP toxic, but it could be equally dangerous in another way. HTP must be kept in absolutely pure tanks and fuel lines, otherwise it will decompose or, if mixed with particular impurities, would explode. HTP was later used to fuel the torpedoes on the Russian submarine *Kursk*, with disastrous results when they exploded in August 2000.

Hydrogen engines for the moon landings

	11D54	11D56	11D57
Use on N1-L3M	3rd stage	Block R	Block S
Design body	OKB-165	OKB-2	OKB-165
Designer	Lyulka	Isayev	Lyulka
Vacuum thrust tonnes	40	7.3	40
Pressure (atmospheres)	60	60	60
Specific impulse	445	461	456
Burn time (sec)	570	800	800
Number of re-starts possible	—	5	11
Weight (kg)	656	292	750

Source: Varfolomeyev (1995–2000)

FOURTH FLIGHT OF THE N-1

Thus, by 1972 the N-1 was in redesign. The old N1-L3 plan had now been superseded by a more ambitious plan, using a redesigned launcher, equipped with the hydrogen-powered engines that had brought the Americans so much success. Once the N-1 was perfected, Mishin could look forward to eclipsing the Apollo landings with the beginnings of a Soviet base on the moon. The Americans might get there first, but Soviet cosmonauts would be the first to really *live* there and explore. Three more N-1s were under construction and one was now almost ready.

With the N1-L3M plan now over a year under way, it was time for the fourth full test of the N-1. Although there had been some pressure to cancel the programme after the third failure, there was a strong conviction that the rocket must now be near to success. Following the third failure, further modifications of the rocket took place:

• Improved aerodynamics, with reduced diameter down from 17 m to 15.8 m.
• Four new vernier engines to improve roll control.

- Better thermal protection for tanks, cables and pipes.
- New control system.
- Improved performance monitoring, with 13,000 sensors sending back data.

This N-1 was the first Soviet launch to use a digital guidance and control system, one overseeing the engines, gyroscopes and accelerometers. The S-530 computer was developed by the Pilyugin design bureau and was used not only for the N-1 but the LOK and LK. The rocket's telemetry system relayed back high-density data, some analysts estimating at a rate of 9.6 Gbyte/sec on up to 320,000 channels on 14 frequencies, so fast that eavesdropping American electronic intelligence satellites could not keep up. Commands could be sent up to the ascending N-1 at the same pace.

Much improved engines were also in preparation, to be installed on the fifth flight model. The fourth N-1 launch took place on 23rd November 1972, directed by Boris Chertok, Mishin's deputy (the chief designer was in hospital at the time). This N-1 carried a dummy LK but, for the first time, a real LOK, which was intended to be put in lunar orbit and return to Earth. To reach the moon, the N-1 would have used a southbound course for translunar injection. A flight plan was approved by Vasili Mishin in July and subsequently published, highlights of which were:

- Burn-out of Earth orbit after one day on 24th November.
- Two course corrections *en route* to the moon.
- Lunar orbit insertion after 98 hours, orbiting at 175 km on 28th November.
- Change in lunar path on the 5th and 27th orbit.
- Descent to 40 km over the moon.
- Landing site photography on orbits 14, 17, 34 and 36.
- Jettison the dummy LK on orbit 37.
- Drop the LOK orbital module on orbit 39.
- LOK to blast back to Earth on orbit 42, on 1st December.
- Course corrections 24 hours after trans-Earth injection and 6 hours before reentry.
- Splashdown in the Indian Ocean on 4th December (Clark, 2002).

At 72 sec after take-off, the fourth N-1 was flying longer, higher and faster than any of its predecessors. Hopes rose that the first staging of an N-1 might now take place just short of the 2 min mark. At 90 sec another hurdle was passed when the six core engines were shut down on schedule (this was a procedure to reduce G forces and vibrations). Then all of a sudden it all went wrong again. Engine #4 caught fire, for reasons that were never satisfactorily explained, right at the end of its burn. There was then the bright flame of an explosion at the tail. The rest of the rocket then quickly blew apart, mere seconds from second-stage ignition. The escape rocket engine fired the payload, the LOK, free. Again, a human crew would have survived.

Higher and faster, the fourth flight of the N-1

Flights of the N-1

Date	Outcome	Payload
21 Feb 1969	Failed after 70 sec	Real L-1S, dummy LK
3 Jul 1969	Failed after 6 sec	Real L-1S, dummy LK
27 Jun 1971	Failed after 51 sec	Mockup LK, mockup LOK
23 Nov 1972	Failed after 107 sec	Real LOK, dummy LK
Aug 1974	Cancelled	Real LOK, LK, block D

THE PLOTTERS MOVE: CANCELLATION OF THE N1-L3M PROGRAMME

The engineers again set to work to tame this difficult beast. The volume of telemetry received probably assisted them greatly. This time, the following changes were introduced:

- New, much improved engines.
- Improved protection for propellant lines.
- Improvements to the fire extinguisher system.
- Faster performance of KORD.

Two new N-1s were built, the first set for launch in August 1974 and the second later that year, with the intention of making the N-1 operational by 1976 and then proceeding to the L-3M plan straight thereafter. A further four N-1s were at an advanced stage of construction and four more were being built. The flight plan was similar to the fourth mission, but this time a functioning LK would be carried. All the manoeuvres short of a lunar landing would be carried out and the LOK would return to Earth after four days of orbiting the moon. Again, hopes began to rise. Assuming the fifth and sixth flights were successful, the seventh would be manned.

Manned lunar flights were not the only missions scheduled. Approval was given for a large, 20-tonne spaceship to be sent to Mars using the N-1 on a sample return mission. This was called Project 5M, led by Sergei Kryukov, later director of OKB

Lavochkin. A date was even set for the launch: 17th September 1975, with a landing on Mars on 22nd September 1976, liftoff from Mars on 27th July 1977 and a return to Earth on 14th May 1978 [19].

Mishin was close to bringing the fifth N-1 down to the pad in May 1974. It was scheduled to fly in August 1974, with the fully improved Kuznetsov engines. An all-up unmanned mission in lunar orbit was scheduled. Even as he did so, the plotters moved.

Mishin had come increasingly under fire not only for the failures of the N-1 programme but also for the difficulties experienced in other parts of the programme. The early 1970s were bad years for the Soviet space programme, for not only were there the problems with the moon programme, but three cosmonauts were lost on Soyuz 11, three space stations were lost over 1972–3 and a fleet of four probes sent to Mars in 1973 suffered a series of computer failures. In some senses, it is a surprise that he lasted as long as he did. Mishin was aware of the criticism, but not that his enemies were preparing to move against him, which they did when he was in hospital. In May 1974, they persuaded Leonid Brezhnev to remove Vasili Mishin from his post as chief designer. He was dismissed on 15th May and replaced at once by Chief Engine Designer Valentin Glushko, who was shortly elevated to membership of the Central Committee of the Communist Party, the apex of political power.

Within days, Glushko suspended the N-1 programme. The sudden suspension of the programme cause widespread shock throughout the Soviet space programme, most so in Kyubyshev where it was built. Alexei Leonov recalls what a devastating blow this was. He blamed Mishin for his failure to present his case properly to the political leadership [20]. Leonov believed that, had Korolev lived, the Soviet Union would certainly have sent a cosmonaut around the moon first. He was less sure that they could have landed on the moon first, but Korolev could have learned from the mistakes with the N-1. They would have got there in the end.

Even after Mishin had been removed, his engineers lobbied hard to be permitted even suborbital flights down the Tyuratam missile range, but to no avail [21]. Others argued, equally unsuccessfully, that even if the moon programme were to be abandoned, the N-1 would still be needed to launch large space stations. Some took out the old N-1-for-Mars design, now called the N-1M, trying to reinvent the rocket for its original mission, intended as far back as 1956. This inevitably prompted a rival design from Chelomei, the UR-700M and then the UR-900, raising the tedious prospect of the battles to the moon being refought again, but this time all the way to Mars [22].

Over the next two years, the Soviet space programme was gradually reoriented, but in a much more fundamental way. The future of the space programme was fought out at a meeting of the Military Industrial Commission on 13th August 1974 [23]. The main imperative seems to have been Glushko's desire for a clean sweep, replacing the N-1 with his own family of launch vehicles (ultimately this evolved into Energiya); a reaffirmation of the value of orbital stations, where the USSR had achieved some modest success; and the need for a space shuttle to match the Americans. The military were not interested in going to the moon, but they were interested to match the shuttle. Now that he had finally triumphed over his dead rival, Korolev and his still alive successor Mishin, Glushko very much wanted to remake the Soviet space programme in his own image [24]. Glushko was undoubtedly a brilliant engineer, but critics found

Valentin Glushko, now chief designer

him petty, gossipy, vainglorious and someone who liked to settle old scores. The political leadership was anxious to reign back costs and even Brezhnev, a supposed lover of *projets de grandeur*, understood the enormous cost to the Soviet economy of moon programmes. Although large-scale lunar and Martian projects continued on the drawing board for another two years, enthusiasm for them diminished to the point that they could be finally buried. Again, the Soviet decision-making process moved slowly and, apart from suspending the N-1, nothing was decided immediately. A consensus emerged, driven by Glushko, who had now combined his old bureau, OKB-456, with Korolev's old OKB-1, not to mention the Kyubyshev plant as well, to form the greatest mega-bureau of all time, Energiya. Following his death in 1989, they were again separated, the former becoming RKK Energiya and the later Energomash.

The three great chief designers of the Soviet space programme

1946–66	Sergei Korolev
1966–74	Vasili Mishin
1974–89	Valentin Glushko

In March 1976, the N-1 was finally cancelled and the order was given to destroy all the N-1 hardware. Project 5M to Mars was cancelled, though the absence of the N-1 was not the only reason (it was eventually recognized as being over-ambitious). The only items to survive were: the NK-33 rocket engines, which were stored away in a shed in the Kuznetsov plant in Kyubyshev; four lunar landers, now to be found in various museums; and half an N-1 fuel tank, which was converted to a bandstand shelter in a park in Leninsk. The N-1 pads were converted to serve for Glushko's new rocket, the Energiya launcher, and it was from one of them that his *Buran* space shuttle made its first and only mission in November 1988. As for the former chief designer, Mishin was

sent to lecture at the Moscow Aviation Institute, and, when *glasnost* broke, emerged
to break the story of the N-1.

Heartbreaking though these decisions were for the designers, the cosmonauts who
had hoped to fly to to the moon also felt an acute sense of disappointment. What
happened to the cosmonaut squad? Once Apollo 8 had flown around the moon, the
prospects of an L-1 manned mission around the moon receded, although briefly
rekindled when consideration was given to a mission to mark Lenin's centenary in
1970. With the failure of the second N-1 rocket in July 1969 and the American landing
on the moon later that month, the prospects of a Soviet manned flight to the moon
depended on the taming of the N-1 rocket, which was nowhere in sight. The squad's
members had so little to do that they were permitted to make overseas trips, though
some were recalled when they told too much of Soviet intentions. Autumn 1969 saw
the *troika* flight of three Soyuz spacecraft, mainly taking cosmonauts from the main
Soyuz training groups but also some less prominent members of the lunar group (e.g.,
Vladislav Volkov). Plans were put forward for 1970–1 for at least one set of *Kontakt*
missions to test out the lunar orbit docking system, with members drawn from the
moon teams and farther afield. These missions were eventually cancelled in late 1970.
When the head of the cosmonaut squad, General Kamanin, came to assemble his
crews for the first manned space station missions in spring 1971, he chose cosmonauts
from the round-the-moon and lunar-landing teams, like Nikolai Rukhavishnikov
(research engineer, Soyuz 10) and Alexei Leonov (original commander, Soyuz 11).
No specific training was ever done for the N1-L3M missions and no simulators were
ever built. The moon team was formally disbanded in May 1974, matching the
suspension of the N-1 programme, although there cannot have been many left at
this stage, most having been reassigned to the manned space station programme. Last
to go was navigator scientist Valentin Yershov, who alleged he was put out either for
not joining the party or else to make way for nominees of new Chief Designer Valentin
Glushko.

Winding down the moon race: cutbacks and redirection

1 Jan 1969	Party and government resolution to continue the moon programme, develop an unmanned alternative programme and develop space stations.
Sept. 1969	First plans drawn for the N1-L3M.
Spring 1971	N1-L3M presented to expert commission.
August 1971	Cancellation of N1-L3 programme, replaced by the N1-L3M programme.
15 May 1972	*Technical proposals for the creation of the N1-L3M complex* approved.
May 1974	Mishin deposed; Glushko becomes chief designer; N-1 suspended. Cosmonaut members disbanded
March 1976	N-1 finally cancelled.

A REAL ALTERNATIVE: SPACE STATIONS

The government decision of 8th January 1969 to reorientate the space programme
around unmanned lunar exploration and space stations did lead to a successful

Salyut – an alternative programme

programme of unmanned lunar exploration (*Chapter 7: Samplers, rovers and orbiters*). More important in the long term, it propelled the Soviet Union into becoming the leading country in the development of space stations. In January 1969, at the reception for the Soyuz 4–5 cosmonauts, Mstislav Keldysh had announced that space stations would be the main line of development of the programme. In October 1969, advocates of the programme cunningly slipped into a speech by Leonid Brezhnev the declaration that they were now *the* main line of development.

Space stations had always featured in Soviet space thinking, back to the time of Tsiolkovsky. Korolev had brought forward outline designs of a manned space station in the early 1950s, to be launched by his N-1 rocket. The first space station programme to win approval for development was a proposal put forward by Vladimir Chelomei's OKB-52 and approved as the Almaz military space station programme (1964). This complex design made slow progress and by 1969 was still some two or three years distant. Accordingly, the decision was taken in early 1970 to combine the Almaz design of OKB-52 with Soyuz hardware developed by OKB-1 so as to construct a space station as soon as possible. Despite these hasty and makeshift origins, not to mention Chelomei's opposition, the space station was actually built quite quickly, in only a year. It was launched in 1971 as Salyut and duly became the first manned orbiting space station, a full two years before America's Skylab. As the moon programme encountered ever more difficulties, the Russians gave their space station programme ever more retrospective justification.

The Soviet Union's space station programme in the early to mid-1970s was cursed with difficulties, and these certainly contributed to Mishin's downfall as much as the moon race. The first crew to reach Salyut was able to link to the station, but not dock properly or enter the station and had to make an emergency return. In the worst-ever disaster to affect the manned programme, the three cosmonauts who flew to Salyut in June 1971 perished on their return, when a depressurization valve opened at high altitude. The next Salyut crashed on launch in July 1972.

The racing days of the moon programme echoed again in spring 1973 as the Americans at last prepared to launch their space station, Skylab. To match Skylab, the Russians prepared two space stations for launch, planning to have both of them operational and occupied at the same time. The first, which was also the first Almaz station, suffered an on-orbit engine explosion and had to be abandoned before it was manned. The second accidentally exhausted all its fuel on its first orbit and also had to be abandoned. It must have galled the Russians that America's Skylab then went on to become such a stunning success.

The first successful space station occupation did not take place until July 1974, when Pavel Popovich and Yuri Artyukin occupied the second Almaz station. This was the first flight after the dismissal of Mishin and the first to take place on Glushko's watch. Even then, the space station programme was to suffer many setbacks and disappointments. Soyuz 15, 23 and 25 had to come down early when their docking manoeuvres failed. The space station programme did not reach maturity until Salyut 6 (occupied 1977–82) and 7 (occupied 1982–6). Here, Soviet cosmonauts learned to live in space, pushing back the frontiers of long-distance flight to 96, 139, 175, 185, 211 and 237 days. Salyuts received regular visitors: unmanned refuelling craft and visiting missions from the socialist countries. With Mir (occupied 1986–2000), the Soviet Union built a permanent orbital station. Mir became to the Soviet programme what Apollo had been to the Americans. Only by returning to its roots in the writings of Tsiolkovsky and the other early visionaries did the programme at last find its true vocation.

THE RACE THAT NEVER WAS

Conventional wisdom about why the Russians lost the moon race is that their technology was inferior and simply could not match the sophistication of Apollo. During the 1970s and 1980s, most Western observers took the view that the Soviet Union never had the technical capacity to send cosmonauts to the moon or land them on it.

Examination of the two paths taken to the moon by the space superpowers shows that this is not the case. The Soviet Union:

- Proved, with Zonds 7 and 8, that it could send cosmonauts around the moon and recover them safely, using different return trajectories.

- Successfully tested out its lunar lander in Earth orbit (T2K: Cosmos 379, 398, 434).
- Built and flew a manned spacecraft for the moon mission that continues to fly to the present day (Soyuz).
- Tested out the key manoeuvres for landing on the moon (Cosmos 382).
- Built high-performance first-stage rocket engines, the RD-270; and the hydrogen-powered upper stages developed by the Lyulka and Isayev bureaux.
- Flew a sophisticated programme of unmanned lunar exploration, with three sample return missions (Luna 16, 20, 24), two rovers (Lunokhod 1, 2) and two orbiters (Luna 19, 22).
- Developed and tested (Soyuz 4/5) a successful spacesuit, now the *Orlan*.
- Built a worldwide land- and sea-based tracking network.
- Pioneered the sophisticated high-speed skip reentry technique.

It is true that the N-1 was not flown successfully. However, the balance of probability is that it would have flown successfully in 1974. The N-1 was the first rocket to have a fully digital computer control system, far ahead of its time. The engine developed for the N-1, the NK-33, was tested in the 1990s and shown to be one of the best in the world, not just then but 30 years later. There are very few rocket programmes where a rocket has not been eventually tamed. Ironically, one of the few others was Europa, cancelled at almost the same time as the N-1 (April 1973) after six failures in a row. Likewise, the roots of that cancellation were political rather than technical.

Students of history therefore cannot explain the Soviet failure in terms of techno-logical shortcomings alone, but must look deeper. The failure of the Soviet Union to reach the moon was, at its heart, a political and organizational failure, not a technical one. Writing about these events years later, Chief Designer Vasili Mishin blamed under-investment, lack of financial control, the dispersal of effort between design bureaux and poor management of the 26 government departments and 500 enterprises involved. The investment was only 2.9bn roubles or $4.5bn compared with Apollo's $24bn. 'They underestimated the technical difficulties involved and should have done ground-testing,' he said.

These judgements, although some might criticize them as self-serving, probably come quite close to the mark, and it may be useful to deal with each in turn. First, the Soviet Union probably had only half the national resources to draw on in mounting a moon expedition than those of the United States. Throughout the moon race, the Soviet Union's gross national product (GNP) was about half that of the United States. In 1957, the United States' GNP was $450bn, the Soviet Union's $210bn, 46.6% of the former. In 1969, the year of the moon landing, the United States' GNP was $930bn, the Soviet Union's $407bn, a slight relative disimprovement at 43.7% [25]. Even if the Soviet proportion of GNP spent on space was more, it was still much less than the American spending, on a dollar-for-dollar basis. American estimates are that the USSR spent about 1.25% of its GNP on the space programme during its peak years, 1966–70. Central Intelligence Agency estimates are that Russian spending rose from $1bn in 1962 to $5bn in 1966, levelling off at $5.5bn during the peak of the moon race

[26]. Of this, the N-1 accounted for about 20% of spending, or $4.8bn (quite close to Mishin's figure).

A lower rate of spending was not necessarily an overwhelming problem, if those smaller resources had been very carefully spent. During the early days of the space programme, later and romantically called 'the golden years', the Soviet Union had clearly punched far above its weight through the astute deployment of limited resources. From the early 1960s, the Soviet Union began to squander its limited resources. The decision of 1964 authorized not one, but two moon programmes, the N-1 and Chelomei's UR-500K. It was actually much worse than that, for by the mid-1960s Russia was not only running two moon programmes, but – if space station, spaceplane and other military programmes are taken into account – no fewer than *seven manned* space programmes *at the same time*. As so many of these programmes were being run by different bureaux, few economies of scale could be achieved. The dispersal and duplication of energies was something which the Soviet economy could afford even less than the American.

The squandering of resources was exacerbated by the rivalry of the different design bureaux and the inability of the Soviet political system to cope with them. Whilst Western analysts imagined a space programme run by a centrally directed command system in which orders were given and bureaux snapped to attention, the opposite was the case, with rival bureaux relentlessly seeking the patronage and support of networks of party and government coalitions. Not only that, but the warring factions constantly sought to have decisions revised and remade, like the UK-700 project which managed to get back on the agenda several times after it was supposedly killed off. The command economy was unable to overcome these problems and command its participants to work effectively together. The spectacle of Khrushchev trying to get his designers Korolev and Glushko into his dacha to make peace – and failing – was one never contemplated in Western understandings of how the Soviet system worked. Although its effects can never be measured, the diversion of energies into such rivalry must have exerted a huge toll on the programme.

A further political failure was the gross misjudgement of American intentions. There is no doubt that the Soviet Union failed to appreciate the significance of President Kennedy's speech in May 1961. The documentary record shows that its implications only began to dawn on the Soviet decision-making system from mid-1963 onward. Even then, the Soviet decision to go to the moon was not made until August 1964, three years after the American one. The actual method was not confirmed until the meeting of the Keldysh commission in November 1966 and the subsequent government decision of February 1967 – when the Americans were less than 30 months away from a landing. It was ironic that the Russians, who had provoked the Americans to competing in a moon race, realized too late that there was a real race under way.

There was one particular misjudgement for which it would be harder to fault them. The American decision to move up Apollo 8 for a moon-orbiting mission clearly took the Russians aback. Contrary to Western notions of Russian recklessness with human lives, they took a cautious approach, insisting on four successful around-the-moon missions before they would put a cosmonaut on board. Yet here the Americans decided, in 1968, to send an Apollo into orbit around the moon on the first manned flight of the Saturn V. Although the gamble paid off, it was nevertheless a risky move. Years later, some of those closest to the decision still recoil at just how risky it was [27]. Despite all the failures of Proton and N-1 on launch, it is some consolation that the launch escape system functioned every single time and no cosmonaut would have been lost on launch. Despite pressure from the Kremlin, the people running the space programme never gave serious consideration to rushing a manned Zond around the moon over 7th–9th December 1968, largely because they felt further testing was required. This must have been a difficult decision, but it was the right one.

Two other factors were also important in the outcome of the moon race. The decision to skip intensive ground-testing for the first stage of the N-1 was a bad mistake and ultimately fatal to the programme. However, some comments should be entered in mitigation. Chelomei's Proton did have the benefit of intensive ground-testing but its miserable development history cost the Russians the first around-the-moon flight and squandered countless payloads. Korolev probably calculated that getting the ground-testing systems for the N-1 built would delay him at least a further year, possibly two, and this was time and money he did not have. Better to take a calculated risk that the problems could be overcome quickly enough, as they had been in the past. In reality, all rocket designers seem to have underestimated the problems of integrating powerful rocket engines and they continued to so do for many years. The development histories of the N-1 and UR-500K were not exceptional: but the Saturn V was.

The second factor was of course Korolev. His loss came at a crucial juncture in the moon race. The way in which he held the Soviet space programme together in its early years and his ability to organize people, bureaux, politicians and talent was legendary. The N-1 could never have got as far as it did without Korolev. The verdict of most of those who knew him was that – with Korolev – the Soviet Union might well have gone around the moon first. The USSR would probably not have landed on the moon first, but he would have given the Americans 'a darn good run for their money'. All agree that he was the only person who could have pulled it off. Mishin, by his own admission, was never able to tame the other design bureaux the way Korolev did. Mishin: 'If Korolev had lived, we would have made more progress.' Though capable in his own ways, he lacked the same drive, organizational ability, relentlessness or capacity to knock heads together. Valentin Glushko had an ambition to match Korolev, but was less able to manage his political masters and flawed by a preparedness to settle scores rather than see projects on their merits. The chief designer system, which served the Soviet Union so well in some respects, was ultimately less successful than American teamwork and its clinical division between political and administrative leadership.

At its heart, the Soviet Union lost the moon race because it misjudged American intentions and resources, mobilized its fewer resources too late and failed to control its competing empires of designers and rocket-builders. Ironically, the Americans won the moon race by showing that they could professionally run a rigorously managed, state-led programme of the type that the Russians were supposed to have but which we now know they did not. Whatever the causes, the winning of the moon race by the

Americans may have had profound political consequences. Despite the Vietnam War, despite many domestic difficulties, the United States reasserted itself, through the moon landing, as the leading technological nation in the world. To John F. Kennedy, this had been the imperative of his era. Kennedy had taken the view that if the United States were to lead what he called the free world, it must prove that it was more capable than its rival. The developing countries, especially, looked to whichever country would be most successful in the mastery of space. The United States went on under the subsequent presidency of Ronald Reagan to rise to a military dominance to become, by the 21st century, the only superpower in a unipolar world. Did the moon victory play a part in this?

By contrast, the loss of the moon race became, in the eyes of subsequent historians, a symbol of the Brezhnev period (1964–82), formally labelled during the time of *perestroika* 'the years of stagnation and decline'. During the period 1957–64, Nikita Khrushchev was able to portray the Soviet Union abroad as an energetic, socially progressive, even liberalizing country able to demonstrate how state-led planning and space-led investment could be an instrument for modernization. Yuri Gagarin's flight became, in the broad canvass of the Soviet years 1917–91, the absolute zenith point of the communist project. But what if Alexei Leonov *had* been first to step upon the moon? This was an interesting exercise explored by the British Broadcasting Corporation [28]. The moon landing might well have given the Soviet system a new lease of life, a new military and political confidence. The Russians might have gone on to establish lunar bases (*Brezhnevgrad*?) and carry out the Mars missions originally projected by Tikhonravov and Korolev in the 1950s, bringing the hammer and sickle with them. How the Americans would have responded is difficult to predict. Unlike the case in the 1950s, they would have lost a contest in which they had specifically set down the goals. Various scenarios are possible, but it is much less easy to see the United States as the unchallenged empire it subsequently became. The moon landing may indeed have been the crucial turning point in 20th century history.

REFERENCES

[1] Gorin, Peter A.: Rising from the cradle – Soviet public perceptions of space flight before Sputnik, in Roger Launius, John Logsdon and Robert Smith (eds): *Reconsidering Sputnik – forty years since the Soviet satellite*. Harwood Academic, Amsterdam, 2000.

[2] Raushenbakh, Boris: The Soviet programme of moon surface exploration, 1966–79. American Astronautical Society, *History* series, vol. 23, 1994.

[3] Clark, Phil: Analysis of Soviet lunar missions. *Space Chronicle, Journal of the British Interplanetary Society*, vol. 57, supplement 1, 2004.

[4] Sokolov, Oleg: The race to the moon – a look back from Baikonour. American Astronautical Society, *History* series, vol. 23, 1994.

[5] Oberg, Jim: Soyuz 5's flaming return – cosmonaut survives a reversed reentry. *Flight Journal*, June 2002, vol. 7, #3.

[6] Clark, Phil: Analysis of Soviet lunar missions. *Space Chronicle, Journal of the British Interplanetary Society*, vol. 57, supplement 1, 2004.

[7] Gracieux, Serge: Le joker Soviétique. *Ciel et Espace*, mai/juin 2005. For a Russian account of the mission, see N.K. Lontratov: 'Late' lunar soil. *Novosti Kosmonautiki*, vol. 15, 1994.

[8] Clark, Phillip S.: Masses of Soviet Luna spacecraft. *Space Chronicle, Journal of the British Interplanetary Society*, vol. 58, supplement 2, 2005.

[9] Sokolov, Oleg: The race to the moon – a look back from Baikonour. American Astronautical Society, *History* series, vol. 23, 1994.

[10] Sagdeev, Roald Z.: *The making of a Soviet scientist*. John Wiley & Sons, New York, 1994.

[11] Clark, Phil: Analysis of Soviet lunar missions. *Space Chronicle, Journal of the British Interplanetary Society*, vol. 57, supplement 1, 2004.

[12] Young, Steven: Soviet Union was far behind in 1960s moon race. *Spaceflight*, vol. 32, #1, January 1991.

[13] Mitchell, Don P. (2003–4):
   – Soviet interplanetary propulsion systems;
   – Inventing the interplanetary probe;
   – Soviet space cameras;
   – Soviet telemetry systems;
   – Remote scientific sensors, *http://www.mentallandscape.com*

[14] Minikin, S.N. and Ulubekov, A.T.: *Earth–space–moon*. Mashinostroeniye Press, Moscow, 1972.

[15] Pesavento, Peter and Vick, Charles P.: The moon race end game – a new assessment of Soviet crewed aspirations. *Quest*, vol. 11, #1, #2, 2004 (in two parts).

[16] Mosnews: Soviet scientists planned 'invulnerable' military headquarters on the moon. *Mosnews*, 20th September 2004.

[17] Wade, Mark: *Energiya – the decision, http://www.astronautix.com*, 2000

[18] Lardier, Christian: Les moteurs secrets de NPO Energomach. *Air & Cosmos*, #1941, 18 juin 2004.

[19] Lardier, Christian: En route vers Mars – et de trois! *Air and Cosmos*, #1895, 20 juin 2003.

[20] Leonov, Alexei and Scott, David: *Two sides of the moon – our story of the cold war space race*. Simon & Schuster, London, 2004.

[21] Clark, Phillip S.: The Soviet manned lunar programme and its legacy. *Space policy*, August 1991.

[22] Zak, Anatoli: *The UR-700 M launch vehicle, http://www.russianspaceweb.com*

[23] Wade, Mark: *Energiya – the decision, http://www.astronautix.com*, 2000.

[24] Sagdeev, Roald Z.: *The making of a Soviet scientist*. John Wiley & Sons, New York, 1994.

[25] Stoiko, Michael (1970): *Soviet rocketry – the first decade of achievement*. David & Charles, Newton Abbot, UK.

[26] Pesavento, Peter: Soviet space programme – CIA documents reveal new historical information. *Spaceflight*, vol. 35, #7, July 1993.

[27] Andrew Smith: *Moondust – in search of the men who fell to Earth*. Bloomsbury, London, 2005.

[28] British Broadcasting Corporation: *What if?* Broadcast, Radio 4, 3rd April 2003.

7

Samplers, rovers and orbiters

When Luna 15 was smashed to pieces in the Sea of Crises in July 1969, Russia's plan to upstage Apollo by the first automatic recovery of lunar soil came unstuck. But the Soviet Union permitted the programme to continue, for two reasons: first, because the series could produce a credible automatic programme for the exploration of the moon; and, second, because the series was important if the Soviet man-on-the-moon programme were to be completed after all. Such hopes still existed in reality up to the summer of 1974 and on paper for another two years.

Luna 15 was the first of the Ye-8-5 soil sampler missions to leave the Earth, one earlier mission having failed on launch. Some considerable work was still required for such a mission to be successful. A number of lessons had arisen from the troubled experience of Luna 15. There had been considerable difficulties controlling the craft. Luna 15's original orbit had been far from that intended. The radar had presented problems. Despite delaying the final landing manoeuvre, the final burn had not proved to be sufficiently precise. In the months that followed the loss of Luna 15, the Lavochkin engineers made the adjustments that they felt sure could guarantee success the next time.

The Lavochkin engineers were convinced that the basic design was sound. Although the three missions had been launched hastily, the basic Ye-8 design, originally intended for lunar rovers, had a lengthy and careful design over many earlier years. The sample return spacecraft consisted of three parts: a descent stage, ascent stage and return cabin.

Ye-8-5 lunar sample return spacecraft

Height	3.96 m
Weight on launch	5,750 kg
on moon	1,880 kg

KT descent stage
Engine   One 11D417

Ascent stage
Weight   520 kg
   incl. propellant   245 kg
Height   2 m
Thrust   1.92 tonnes
Propellant   Nitric acid and UDMH

Return cabin
Weight   39 kg
Diameter   50 cm

For radio communications, the lander carried a cone-shaped antenna on a long boom, working on 922 MHz and 768 MHz. Uplink was received on 115 MHz. A dish-shaped radar was located on the bottom of the spacecraft.

The Lavochkin bureau still had another three sample return spacecraft available. All were duly launched in the period following the Apollo 11 landing, on 23rd September 1969 (Cosmos 300), 22nd October 1969 (Cosmos 305) and 6th February 1970. On Cosmos 300, there was a leak in the oxidizer tank of block D, which depleted the entire supply during Earth orbit injection and could not fire out to the moon, leaving the spacecraft to crash back to Earth four days later. On Cosmos 305, the attitude control system failed and block D did not get into the right attitude to fire to the moon, crashing back near Australia. With the February 1970 launching, Proton's first-stage engines were erroneously turned off at 127 sec and it cratered downrange.

Of the first five attempts, only one had left Earth orbit. A second batch of sample return spacecraft was now constructed and there was a delay until the first of the new spacecraft could be available. In the meantime, concerted efforts were applied to attempt to fix the appalling record of the Proton rocket. Proton's unreliability had not only cost the moon programme dearly, but dogged the interplanetary pro-gramme, destroying two of a new series of Mars probes in March 1969. Eventually, Georgi Babakin persuaded the minister responsible for the space programme, Sergei Afanasayev, to introduce a requalification programme. This took place over spring and summer 1970, culminating in a suborbital test on 18th August 1970. This led to a swift and radical improvement in performance, but Soviet space histories might have been happier, had these changes been introduced sooner.

LUNA 16

The first of the new batch of Ye-8-5 spacecraft was not available until the following month. Luna 16 was launched on 12th September 1970, and it headed out moonwards on a slow four-day coast. In contrast to the great media interest which Luna 15 had attracted, Luna 16 went virtually unremarked by the Western media. This was a pity,

Luna 16, testing before launch

for Luna 16 was a remarkable technical achievement by any standard. Its flight coincided with what became known to the world as Black September. Four airliners were seized in the space of a few hours by Palestinian fighters; the aircraft were hijacked to a remote airstrip called Dawson's Field in Jordan; King Hussein's army moved in to crush the Palestinians. The world looked on, mesmerized.

Luna 16 carried, like Luna 15 before it, the new KTDU-417 main engine built by the Isayev design bureau. The KTDU-417 had a throttleable engine ranging from

750 kg to 1,920 kg. At highest thrust, it had a specific impulse of 310 sec, able to burn for 10 min 50 sec using up to four tonnes of propellant. This engine was built to perform mid-course correction, lunar orbit insertion, pre-descent burn, the 'dead-stop' burn to take it out of lunar orbit and the final burn 600 m above the moon. There was also scope for further manoeuvres in lunar orbit, as had proved necessary on Luna 15 and the engine could be fired up to eleven times. It could also be used at lowest thrust with a specific impulse of 250 sec [1]. The engines for the final stages of landing had a thrust of 210 and 350 kg.

Luna 16 burned its engine on the first day for 6.4 sec to make a course correction. Luna 16 entered moon orbit on 17th September at an altitude of 110 km to 119 km, 71°, 1 hr 59 min. The aim was to achieve a circular lunar orbit around 100 km.

After two days, Luna 16 fired its engine to make a 20-m/sec velocity change and brake into an elliptical course of 106 km by 15.1 km, with the perilune over the landing site. Its final path before descent was 15 km by 9 km, so low as to only barely scrape the peaks of the moon's highest mountains. At this stage, the four 75 kg large propellant tanks that had been used for mid-course correction, lunar orbit injection and orbital change were jettisoned.

As Luna 16 skimmed over the eastern highlands of the moon on the 20th, the retrorocket of the 1,880 kg craft blasted and Luna 16 began to fall. First, the main engines blasted for 267 sec, using about 75% of fuel remaining, to kill all forward motion. This was a big burn, 1,700 m/sec. The critical stage had begun. Luna 16 was now over flat lowlands. Sophisticated radar and electronic gear scanned the surface, measuring the distance and the rate of descent. After the 'dead stop' engine burn, Luna 16 was in free fall, coming down at 215 m/sec, until six minutes later it was at 600 m. Then the main engine blasted again. At 20 m, a point detected by Doppler-sounding gamma rays, the retrorocket cut off and small vernier engines came into play. At 2 m, sensing the nearness of the surface, these too cut out, the intention being to achieve a landing speed of 2.5 m/sec or 9 km/hr. Luna 16 dropped silently to the airless surface, bouncing gently on its four landing pads. It was down, safe and sound, on the Sea of Fertility, 100 km from crater Webb. The flat and stony ground was marked only by a few small craters, even if they were not visible during the descent, because Luna 16 had landed in darkness. This and subsequent soil-sampling missions carried stereo cameras of the type carried by Luna 13, so the quality of images should have been very good. The purpose of the cameras was to help to guide the operators of the drill and for such night landings floodlights were carried.

Strong signals were picked up by Western tracking stations. Within hours, the USSR had announced its third soft-landing on the moon – but said no more. The Russians had still not admitted that the intention of the probe was to collect samples.

Meantime, a quarter of a million miles away a 90 cm drill arm swung out from Luna 16 like a dentist's drill on a support. It swung well clear of the base of the spacecraft, free from any area that might have been contaminated by gases of landing engines. The wrist of the drill had a flexibility of 110° elevation, 180° rotation and was able to drill to 35 cm. The drill head bored into the lunar surface at 500 r.p.m. using electric motors for 7 min and then scooped the grains of soil down to 35 cm deep. There it began to hit rock and, rather than risk damaging the drill, the boring was terminated and the sample collected and put into the container attached to the drill head. Like a robot in a backyard assembly shop, the drill head jerked upwards, brought itself alongside the small 39 kg spherical recovery capsule, turned it round and pressed the grains into the sealed cabin, which was then slapped shut.

THE GENIUS OF OKHOTSIMSKY

By the 21st, Luna 16 had spent a full day on the moon. There was still no official indication as to its purpose. Jodrell Bank reported still more strong signals. In fact, what Luna 16 was doing was checking out its exact landing coordinates so as to give the best possible return trajectory. Luna 16 had landed at the lunar equator at 56°E, the perfect place for the direct return to Earth on Dmitri Okhotsimsky's passive trajectory. The return system would now be put to the test.

All was now set for the return of the ascent stage to Earth. The top stage of Luna 16 weighed 520 kg, with the recoverable cabin. There was one engine on the ascent stage, the KRD-61 of the Isayev design bureau. Burning 245 kg of UDMH and nitric oxide, the ascent stage had a specific impulse of 313 sec and could burn just once for 53 sec, sufficient to achieve a velocity of between 2,600 m/sec and 2,700 m/sec. A complication of the 2.9 day return flight was that – to recover the spacecraft in the normal Kazakhstan landing site – liftoff would take place out of sight from Yevpatoria: the moon would be over the Atlantic, where it could be followed by a Soviet tracking ship offshore Cuba.

Twenty-six hours after landing, explosive bolts were fired above the Luna 16 descent stage. On a jet of flame, the upper stage shot off and headed towards the white and blue Earth hanging in the distance. It headed straight up, motor still purring, building up to lunar escape velocity, its radio pouring out details from the four aerials poking out the side. The Sea of Fertility returned to the quiet it had known for eons. The descent stage was the only forlorn reminder of the brief visit. The lower stage on the moon continued to transmit signals for a couple of days until the battery ran out. Only two instruments seem to have been carried: a thermometer and radiation counter.

The returning rocket – capsule, instrument container, fuel tanks and motors – reported back from time to time as it headed for a straight nosedive reentry. These coordinates had to be as precise as possible so as to best predict the landing spot on Earth. At 48,000 km out, the tiny capsule separated from the instrument and rocket package, plunged into the upper atmosphere, glowed red and then white as temperatures rose to 10,000°C as it hit forces of 350 G. Helicopters were already in the air as a parachute ballooned out at 14,500 m. The capsule hit the ground and beacons began sending out a bright beep! beep! signal as rescuers rushed to collect the precious cargo. The mathematicians had done their job well, for Luna 16 came down 30 km from the middle of the intended recovery zone, 80 km southeast of Dzhezhkazgan, Kazakhstan.

Luna 16 stage left on the moon

The small capsule was transferred to a plane and flown at once to Moscow to the Vernadsky Institute of Geological and Analytical Chemistry for analysis. The person in charge of assessing the lunar soil was Valeri Barsukov (1928–92), subsequently to become director of the institute (1976–92). How the scientists ever got the soil container open is a mystery for the entire outer skin of the capsule may well have been welded by the intense heat of the fiery return. Once open, the grey grains of moon dust poured out – loose lumps of dark grey, blackish powder like very dark, wet beach sand. It had small grains at the top and large grains at the bottom where it had begun to encounter rock.

The sample, although small (105 g), provided a considerable amount of scientific information [2]. The following were the main features:

- It was a uniform, unstratified sample.
- Seventy different chemical elements were identified.
- The sample comprised a mixture of powder, fine and coarse grains.
- It had good cohesive qualities, like damp sand.
- The sample was basaltic by character.
- It included some glazed and vitrified glass and metal-like particles.
- The samples had absorbed quantities of solar wind.

The Luna 16 cabin back on Earth

It was a tremendous triumph. The Luna 16 mission had gone perfectly from start to finish. The tricky stages of soft-landing, drilling and take-off were just like the book said they should be. 'It's the decade of the space robot!' heralded the Soviet press. The USSR made great play of how such flights were cheaper than manned flights like Apollo, how they did not expose humans to danger and how versatile space robots could land just about anywhere.

Luna 16 recovery

Luna 16 moonrock

For NASA and Western observers the real significance of Luna 16 lay elsewhere: it confirmed what many, but not all of them, had suspected was Luna 15's real purpose, namely that it was a real challenge to Apollo 11 a year earlier. Russia did have good grounds to celebrate Luna 16. Some of the remarks about its low cost and versatility were exaggerated and Luna 16's sample of 105 g was tiny compared with Apollo, each mission of which brought back well over 20 kg. Luna 16 did not have the same capacity to search around for and select samples as the men of Apollo, for the arm would set the drill into the nearest piece of adjacent surface regardless. The Russians later exchanged 3 g of Luna 16 samples for 3 g each from Apollo 11 and 12. Many years later, the Russians sold 2 mg of soil at Sotheby's in New York, fetching an out-of-the-world price of $442,500. The results of the soil analysis were published in a number of scientific papers over the following years.

DESIGNING A LUNAR ROVER

Although the Lunokhod was portrayed by the Soviet Union as a safer, cheaper alternative to the manned Apollo missions, in fact the Lunokhod long pre-dated Apollo. Originally, Lunokhod was an integral part of the *manned* Russian lunar programme. Moon rovers were to pave the way for manned landings by surveying sites before cosmonauts landed, the L-2 programme. They would leave beacons to guide the LK landing ships in. Later, bigger rovers would be landed and cosmonauts were expected to ride them across the moon (the L-5 programme).

The moon rover was originally designed in Korolev's OKB-1. The preliminary studies were done by Mikhail Tikhonravov in 1960. When the Americans first landed a rover on Mars, the *Sojourner* (1997), it was tiny. By contrast and in typical Soviet style, the Russians started large. Korolev's team determined that the rover should be at least 600 kg, the size of a small car. This would require a launcher much larger than the Molniya then in design, so Korolev made it a candidate for an early version of the N-1 rocket. Korolev issued the order for the construction of a moon rover in March 1963, but the project progressed slowly and was set back when later that year the state Institute for Tractor and Agricultural Machinery Building declined to develop it, deeming the project to be 'impossible'.

So, later in 1963, Sergei Korolev instead turned to VNII-100 Transmash of Leningrad, or the Mobile Vehicle Engineering Unit [3]. In September of that year, Korolev met with VNII Transmash engineers to go through the possibilities.

Alexander Kemurdzhian

Transmash designed tanks for the Red Army – indeed, during the siege of Stalingrad, tanks were sometimes rolled out of the factory straight up to the front line. The important role of Alexander Kemurdzhian in the Soviet lunar programme emerged only in recent years. He was born on 4th October 1921 in Vladikavkaz and entered the Bauman Technological College in Moscow in 1940. When the war broke out, he went to Leningrad Artillery College and participated in some of the epic battles of the war, such as the crossing of the Dniepr. After the war, he worked on truck design, specializing in transmission systems, for which he obtained a doctorate in 1957. Two years later, he moved into the new area of air cushion vehicles (hovercraft). Kemurdzhian had a personal interest in spaceflight (something he made clear to Korolev) and saw the potential for remote-controlled vehicles exploring the planets. The rover project was no sideshow, for in 1964 it won approval – as the L-2 programme – in the 1964 government and party resolution committing the Soviet Union to going to the moon.

The conceptual study was completed in six months, by April 1964. One of the first problems faced by the designers was the load-bearing capacity of the lunar soil, for this would govern chassis, power systems and wheel design. Until such time as soft-landers tested the surface, it would be impossible to know the answer for definite. In an attempt to make the best possible estimate, a conference of lunar and astro-nomical experts were gathered at Kharkov University that year, hosted by Professor Barabashev and also attended by Professor Troitsky of Gorky University and Pro-fessor Sharanov of Leningrad University. In the event, their estimates were broadly

Moon rover on test

correct, being confirmed by Luna 13 two years later. First design sketches were concluded in September 1965.

The rover project was turned over, along with all the other unmanned lunar and interplanetary programmes, to OKB Lavochkin in 1965. Kemurdzhian worked closely with the director of OKB Lavochkin, Georgi Babakin, to finalize what was then called in 1966 the Ye-8. The Ye-8 was originally intended to pave the way for the manned lunar landing. Before the first Ye-8 landed, suitable sites would first be selected by a close-look lunar orbiter. To do this, a version of the rover was adapted for a photography mission in lunar orbit to select a main landing site for the lunar landing, but there was also a reserve one nearby, not more than 5 km distant. Two Ye-8s would then be landed, one at the main site, one at the reserve. These would confirm the suitability of both sites for the manned lunar landing. In an elaboration of the plan, an unmanned LK would be landed near the rover at the reserve site and checked out to see that it was in good working order. If when he landed his LK was disabled, the sole cosmonaut could travel to the reserve LK to return to Earth. In a further version, the cosmonaut could use the rover to travel across the lunar surface from the main site to the reserve site.

A number of designs using different numbers of wheels were considered in the course of 1965–6. The designers considered tractors, walkers and even jumpers, from caterpillar to four-wheel designs. The very first rover design was for a dome carried on four caterpillar wheels, very like a tank. The first rovers were designed to weigh nearly a tonne, about 900 kg. When it was apparent that the N-1 would not become quickly available, the Ye-8 was scaled down so that it could be accommodated within Chelomei's UR-500K Proton. The final rover design was for an unmanned rover. In a further modification of the original plan, the Ye-8 would be launched before the Ye-8LS lunar orbiter, the opposite of what had been intended.

The rover design was settled in 1967 and a 150 kg scaled prototype was constructed in Leningrad that year. A version was tested in the volcanic region of Kamchatka in the Soviet far east, which was the Earth's surface closest in character to the moon. Models were tested in the Crimea and early versions of the transmission gears and wheels were flown out to the moon on Luna 11 and 12 in 1966 and Luna 14 in 1968. Even though it had been scaled back, the final rover was still a substantial piece of engineering. The vehicle, to be called 'Lunokhod' or 'moon walker' in Russian, weighed 756 kg and was 4.42 m long (lid open), 2.15 m in diameter and 1.92 m high. Its wheel base was 2.22 m by 1.6 m. The main container was a pressurized vehicle, looking like an upside down bathtub, carrying cameras, transmitters and scientific instruments. It was kept warm by a small decaying radioisotope of 11 kg of polonium-210. The eight 51 cm diameter wheels were made by the Kharkhov State Bicycle Plant, made of metal with a mesh covering. There was a ninth wheel behind the vehicle to measure distance. Each wheel had its own electric motor. In the event of one wheel becoming completely stuck, a small explosive charge could be fired to sever it. The vehicle was designed to climb slopes of 20° and manage side slopes of 40 to 45°. The main designers were, aside from Kermurdzhian himself, Gary Rogovsky, Pavel Sologub, Valery Gromov, Anatoli Mitskevich and Slava Mishkinjuk.

To guide the route chosen, Lunokhod had four 1.3 kg panoramic cameras similar to those on Luna 16 to scan 360° around the rover and two television cameras to scan forward, with a 50° field of view and 1/25 sec speed. The scan of the panoramic cameras was designed in such a way as to cover the horizon right around to parts of the rover and its wheel base. They provided high-resolution images, $6,000 \times 500$ pixels. Signals were sent back by both an omnidirectional and narrow-beam antenna. The driving camera relayed pictures back to Earth every 20 sec and these enabled a five-person ground crew to drive the Lunokhod: commander, driver, navigator, engineer and radio operator. The rover could go forwards or backwards. Gyroscopes would stop the rover if it appeared to tilt too much forward or backward or to one side.

The selection of the ground crew was an important part of the programme. Two five-man crews were selected from the Missile Defence Corps in 1968 [4]. Volunteers were sent for tests for speed-of-reaction times, short and long-term memory, vision, hearing and capacity for prolonged mental focus and attention. At one stage of their recruitment, they thought they were being trained as cosmonauts. They were under strict instructions not to talk about their work to outsiders. Years later, their names became known. They had been recruited by the Strategic Rocket Forces in the late 1960s when the call had gone out for 'top class military engineers. Young but experienced. Sporting and in a good state of health.' Twenty-five were chosen and sent to Moscow for a special mission, they did not know what. They were put through a series of tests in the Institute for Medical and Biological Problems, where the group was reduced by eight. Then, the seventeen remaining were told that they would be driving machines across the surface of the moon, whereupon three resigned, saying that the responsibility and stress would be too much for them. The fourteen remaining were divided: half were sent off to Leningrad to the VNII-100 design bureau where the Lunokhod was built and the other half were assigned to work on the design with the Lavochkin design bureau. In 1968, construction began of a 'lunardrome' in Simferopol in the Crimea, and the driving teams spent the rest of the year there learning how to drive a Lunokhod model.

Table 7.1. The Lunokhod operators

	First crew	Second crew
Commander	Nikolai Yeremenko	Igor Fyodorov
Driver	Gabdulkay Latypov	Vyacheslav Dovgan
Navigator	Konstantin Davidovsky	Vikentiy Samal
Engineer	Leonid Mosenzov	Albert Kozhevnikov
Radio, antenna	Valeri Sapranov	Nikolai Kozlitin
Reserve	Vasili Chubukin	

Luna 17 descent stage

Lunokhod carried a number of scientific instruments: a French-built 3.7 kg laser reflector, designed to measure the precise distance between Earth and the moon; a RIFMA X-ray fluorescent spectrometer to determine the composition of moonrock; an X-ray telescope; a cosmic ray telescope; and a penetrometer. An energetic particle detector was built by Dr Yevgeni Chuchkov of the Theoretical and Applied Physics Divison of the Skobeltsyn Institute of Nuclear Physics of the Moscow State University, calibrated against similar instruments flown on Zond 1 and 3 and the early Mars and Venera probes.

To get Lunokhod onto the lunar surface, the KT stage was used, of the same type as Luna 15 and 16. This was a frame-shaped spacecraft with a toroidal fuel tank; radar; attitude thrusters; 11D417 engine of between 0.75 and 1.92 tonnes of thrust for mid-course correction, lunar orbit insertion and landing; batteries; and communications. The Lunokhod rested atop the descent stage, and – when the moment came – landing ramps would deploy at either side so the rover could descend to the moon at an angle of up to 45°.

Lunokhod

Weight	756 kg
Diameter	2.1 m
Height	1.35 m
Wheel base	2.22 m × 1.6 m
Number of wheels	8
Wheel diameter	51 cm
Speed	0.8 to 2 km/hr
Operation duration	3 months
Ye-8 with KT stage	1,880 kg

Lunokhod instruments

Laser reflector.
RIFMA (Roentgen Isotopic Fluorescent Method of Analysis) X-ray fluorescent spectrometer.
Extra-galactic X-ray telescope.
Cosmic ray background radiation detector.
PrOP (*Pribori Ochenki Prokhodimosti*) penetrometer.
Ultraviolet photometer (Lunokhod 2 only)

Any benefit that was gained by the success of Luna 16 was turned to double advantage just two months later by Luna 17. The sample return, pushed to the back page by the eruption of political violence in the Middle East, had made little public impact. The same could not be said of its successor, put up on 10th November 1970. The spaceship weighed about 5,750 kg.

INTO THE SEA OF RAINS

Luna 17's mission was, at least for its first six days, apparently identical to that of Luna 16 and 15. A four-day coast out to the moon was followed by lunar orbit insertion circular at 85 km, 1 hr 56 min, 141°. On the 16th, the onboard motor lowered the orbit to an altitude of 19 km. Luna 17's target was nearly a hemisphere away from that of Luna 16. The entire western face of the moon is dominated by a huge, dark 'sea' which is called the Ocean of Storms. In its northwest corner is a semi-circular basin, the Sea of Rains.

After only two days in orbit, reflecting the bright sunlight of the setting sun, Luna 17 skimmed in low over the Jura Mountains. The retrorocket fired. Luna 17 came down as the radar checked the landing site. At 600 m, coming down at 255 m/sec, the final main engine burn was made. Down it came, as softly as a parachutist on a wind-free day. By the time it landed, Luna 17 weighed 1,836 kg. The long shadows of the structure stood out starkly toward the darkening east. For two hours, Luna 17 reported back its position. Russia coolly announced its fourth soft-landing on the moon. A return capsule would be fired back to Earth the next day – or so everyone thought.

Lunokhod descending to the moon

Not so. On the upper stage rested the first vehicle designed to explore another world. It had eight wheels, looking like pram wheels, which supported a shiny metallic car, covered by a kettle-style lid. Out of the front peered two goggle-like television eyes. Above them peeped the laser reflector and two aerials. It was an unlikely-looking contraption – on first impression more the outcome of a Jules Verne or H.G. Wells type of sketch rather than a tool of modern moon exploration. But the wheels were ideal for gripping the lunar surface and less prone to failure than caterpillars. The lid could be raised backward to the vertical and then flat behind, exposing solar cells to recharge the batteries in the Sun's rays. The exposed top of the car was a radiator, discharging its electronic and solar-baked heat. There was genius in its simplicity.

The most dangerous part of the vehicle's journey was probably getting off the platform and onto the lunar surface. Two ramps unfolded at each end, so it could travel down either way if one exit were blocked. Still sitting on the landing platform, ground control commanded the dust hoods to fall off the television eyes. A picture came back at once, showing the wheel rims, the ramp down to the flat bright surface and the silhouette of the landing ramps. There was nothing for it but to signal to Lunokhod to go into first gear and roll down the ramp and hope for the best.

So it was that at 6:47 a.m. on the morning of 17th November 1970, carrying the hammer and sickle, a red flag and a portrait of Lenin, the moon vehicle edged its way down the ramp and rumbled 20 m across the lunar surface. Its tracks were the first wheel marks made on another world. Its television cameras showed its every move and at one stage Lunokhod slewed around to film the descent stage which had brought it there. On day 2 it parked itself, not moving at all, lying there so that its lid could soak

Lunokhod tracks across the moon

in solar energy for its batteries. On day 3 it travelled 90 m, 100 m the following day, overcoming a 10° hill. On the fifth day, with lunar night not long off, it closed its lid, settled down 197 m from Luna 17 and shut down its systems for the 14-day lunar night. At this stage, it had travelled a modest 200 m. A nuclear power source would supply enough heat to keep it going till lunar daybreak.

The Soviet – and Western – press took to Lunokhod with an affection normally reserved only for friendly robot television personalities. There was unrestrained admiration for the technical achievement involved, for it was a sophisticated automated exploring machine. *The Times* of London called it 'a remarkable achievement'. 'A major triumph,' said *The Scotsman*. The *Daily Mail*, in a front-page editorial entitled 'Progress on wheels' gave Lunokhod's designers an effusive message of congratulations. It was the main news story for several days.

The control centre for Lunokhod was, like much else in the venture, a scene straight from science fiction. It was located in Simferopol, Crimea, near the big receiving dishes. Five controllers sat in front of television consoles where lunar landscapes were projected on screens. The crew of five worked together like a crew operating a military tank. Signals were relayed to the drivers by the high-gain antenna which had to be locked on Earth continuously. The drivers operated Lunokhod with a control stick with four positions (forward, backward, stop, rotate), and they could make the rover go either of two speeds forward: 800 m/hr or 2 km/hr, or reverse. If the Lunokhod looked like crashing, either drivers or commanders could press a panic button to turn the electric engine off. Any one wheel could be disconnected individually if it got stuck or there were a problem. Lunokhod was designed to cope with obstacles up to 40 cm high or 60 cm wide, but an automatic system would cut the engine out if it began to tilt. Average speed started at 2.3 m/hr but later increased to 4.8 m/hr. All the wheels ran at the same speed and they turned the rover like a tank by running the wheels faster on one side than the other, until the change of direction was achieved – skid-steering [5]. In reality, driving the Lunokhod proved to be quite a lot more difficult than the drivers expected. The drivers realized at once that the cameras were too low down – it was like being a human on all fours rather than upright. The television cameras were able to provide little contrast: the images were too white, and rocks and craters looked deceptively alike [6]. Driving the moonrover was strenuous and during the lunar days the teams alternated 9 hr shifts, catching up on sleep during the lunar nights.

So great was the excitement of the first Lunokhod that journalists, academicians and scientists flooded into mission control, apparently taking up a general invitation to do so by Mstislav Keldysh. Vistors were not supposed to crowd around the drivers, still less talk. But the situation got out of hand, especially when backseat drivers would exclaim: 'He's going to crash into that rock!' or 'Mind that crater!' Between the natural stress, the heat coming out of the televisions and the backseat drivers, the drivers' pulses crept up to 140 and the stress began to tell. Babakin had had enough. 'Everyone out of here!' he ordered and after that special passes were needed to visit the control room and then in a suitable state of humility [7].

Back on the moon, nighttime temperatures plunged to −150°C and stayed at that level a full two weeks. Lunokhod, lid closed, glowing warmly from the heat of its own nuclear radio isotope, rested silently on the Sea of Rains. It was bathed in the ghostly blue light of Earth as the mother planet waxed and waned overhead. Even as it stood there, laser signals were flashed to Lunokhod from the French observatory in the Pic du Midi and from the Semeis Observatory in the Crimea. They struck the 14 cubes of the vehicle's laser reflector and bounced back. As a result, scientists could measure the exact distance from the Earth to the moon to within 18 cm.

To the east of Lunokhod rose a ridge and the sharp rays of dawn crept slowly over its rugged rocks early on 9th December. Had the moonrover survived its two-week hibernation? This was an anxious moment and pulses began to race when the first command was sent to the Sea of Rains to open the lid. Nothing happened. They tried again and this time the rover responded. It raised its leaf-shaped lid and at once began to hum with life. Four panoramic cameras at once sent back striking vistas of the moonscape, full of long shadows as the Sun gradually rose in the sky. After a day recharging, Lunokhod set out once more. The Lunokhod got into big trouble straight

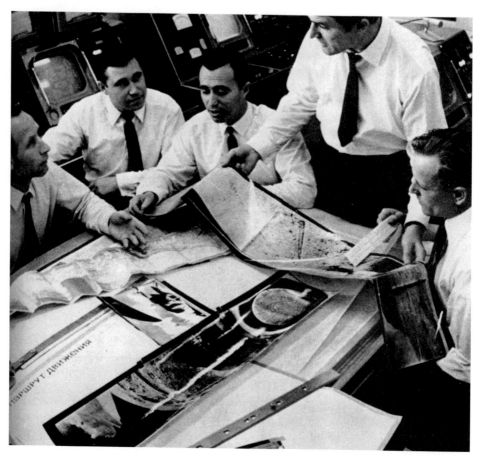

Lunokhod route-planning conference

away. On 10th December, Lunokhod got stuck in a crater and no matter what the drivers did – go forward, go back – it remained stuck. Eventually, after nine exhausting hours, the rover suddenly came free.

The drivers on Earth soon got into their stride and they had the moon car in second gear, swivelling around, reversing and traversing craters and slopes at will. One day it travelled 300 m, more than it had achieved in its first five days in November. Lunokhod took a south–southeast path, skirting around and between craters and parked in December in a crater at the southernmost end of the route, 1,400 m from the landing stage. In January, swivelling round to head back north, the panoramic camera eyes spotted in the distance a range of mountains – the far peaks of the Heraclides Promontory, part of the vast bay encircling the Sea of Rains.

For ground control it was just like being there. From the cosy warmth of their control post they could direct at will a machine a quarter of a million miles away. This prompted romantic notions in the minds of the Earthbound. Radio Moscow promised 'more Lunokhods, faster and with a wider range.' Boris Petrov spoke of mooncars

Lunokhod tracks

that would collect samples and bring them to craft like Luna 16 for transporting home. Others would instal packages on the moon and carry telescopes to the farside where there was radio peace, free from Earthside interference. Other probes would reach the lunar poles.

It turned out that the drivers had been well selected for their mission. The drivers faced several challenges. First, the 20 sec frame transmissions were too slow. Although driving the lunar rover might seem simple enough to a modern generation reared on video games, in reality the crew had to memorize features some distance ahead. The 20 sec time gap between frames meant that Lunokhod could reach a feature – stone, rock, crater, obstacle – a full third of a minute before the crew saw visually that it had arrived. Second, the cameras were set in an awkward place: too low to see far ahead, yet set toward the horizon in such a way as to create a dead zone immediately in front of the rover that the drivers could not see. Third, the light contrasts of the lunar surface made driving difficult, the drivers having to cope with extremes of shadows and glare. Rather than risk driving across shadowless moonscapes, operations were normally halted for two days at lunar high noon. From time to time, Lunokhod would

Lunokhod returns to landing stage

stop to take panoramic pictures. For the drivers, these were good opportunities to orientate the rover and plan the next stage of the journey.

'LUNOKHOD, NOT LUNOSTOP'

Scientists sat in an adjoining room watching the pictures and hearing the comments of the drivers, but were not allowed into the control room. This was quite different from American practice for, when American rovers explored Mars in 1997 (*Sojourner*) and 2004 (*Spirit* and *Opportunity*), the scientists were an integral part of the team. Eventually and Babakin's edict notwithstanding, the principal lunar geologist, Alexander Basilevsky of the Vernadsky Institute for Geochemistry, could bear the separation no longer, brought his chair into the control room and watched quietly from close quarters. There was quite a contrast between the way the Russians approached things on the moon and the way the Americans subsequently did on Mars. Whenever Basilevsky wanted to examine a rock, the drivers wanted to avoid it, for fear of collision or getting stuck – by contrast, the American Mars rovers spent extensive periods getting up close and personal to individual interesting rocks. Georgi Babakin, aware of thirst in *Pravda* for 'how many kilometres did we do today?' once told Basilevsky gently that this was a Luno*khod*, not a Luno*stop*.

As time went on, it became apparent that Lunokhod was not just a playful bathtub on wheels but a sophisticated machine with advanced instrumentation. The soil analyzer RIFMA bombarded the surface with X-rays and enabled ground control to read back the chemical composition of the basalt-type soil. From time to time, the PrOP mechanical rod jabbed into the soil to test its strength. When it did so, it was able to measure resistance. Then, it was turned in the soil, this time to measure turning resistance. Once done, it was retracted and the vehicle moved on. Lunokhod did not only look moonwards: there were two telescopes on board – one to pick up X-rays beyond the galaxy and another to receive cosmic radiation. On 19th Novem-

Lunokhod tracks from landing stage

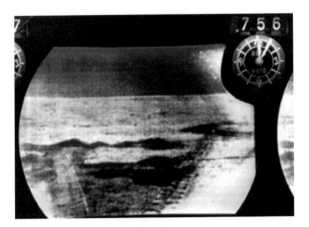

Lunokhod porthole view

ber, Lunokhod recorded a strong solar flare that could have injured cosmonauts had they been on the moon at the time. Lunokhod therefore contained within it several concepts: an exploring roving vehicle; a rock-testing mobile laboratory; and an observatory able to capitalize on the unique air-free low-gravity environment beyond the Earth. The rocks were abundant in aluminium, calcium, silicon, iron, magnesium and titanium. The Sea of Rains had been selected because it was a typical *mare* area.

Come the new year, 1971, Lunokhod was back in action once more, heading back north to its landing site, whence it returned in mid-January. A spectacular photograph of the landing vehicle with ramps and wheel tracks all about reminded the world that Lunokhod was still there, prowling about the waterless sea of the Bay of Rains. The normal procedure was to lift the solar lid when the sun was 5° over the horizon. The first thing Lunokhod would do was radio back its condition (for the record: internal temperature 22°, pressure 753 mm (May)). Now Lunokhod headed north on a much longer journey from which it would not return. By the fourth lunar day, 8th February, scientists were able to compile a map of that part of the Bay of Rains adjacent to Luna 17. On 9th February, the mooncar survived a lunar eclipse when temperatures plunged from +150°C to −100°C and back to +136°C, all in the space of three hours. In March, Lunokhod explored around the rim of a 500 m wide large crater, venturing into smaller craters within the rim of the larger circle.

In April, Lunokhod ventured to a crater field full of boulders over 3 m across. Because of a nearby crater impact, all the black lunar dust had piled up against one side of the boulders as if a hurricane had swept through. Eventually, the geologists persuaded the control room supervisor to concentrate more on science and photographing features of interest and less, as they saw it, on building up distance records for the pages of *Pravda*. Drivers wanted to avoid rocks that might endanger Lunokhod, but the scientists wanted to examine them at close quarters. On 13th April, Lunokhod became badly stuck in loose soil on a crater slope, but by applying full power on all engines it emerged out onto more solid, level ground. This nearly exhausted the system and the rover was parked for the rest of the month, simply

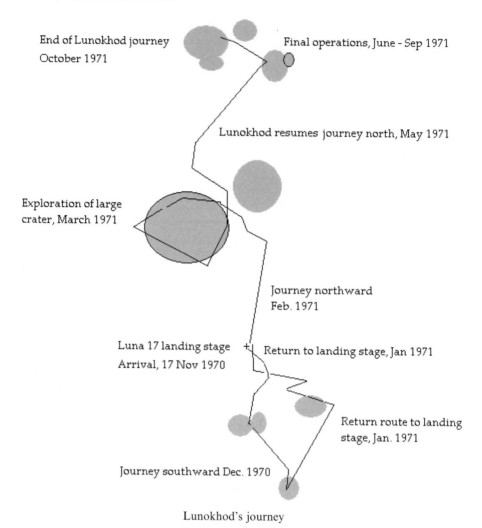

End of Lunokhod journey
October 1971

Final operations, June - Sep 1971

Lunokhod resumes journey north, May 1971

Exploration of large
crater, March 1971

Journey northward
Feb. 1971

Luna 17 landing stage
Arrival, 17 Nov 1970

Return to landing stage, Jan 1971

Return route to landing
stage, Jan. 1971

Journey southward Dec. 1970

Lunokhod's journey

recharging. Lunokhod travelled only 197 m in May, concentrating on static experiments. The mooncar had appeared to be losing power and it was probably decided to concentrate on less energy-demanding experiments (Lunokhod was never expected to last more than six months).

Measurement of the strength of the lunar surface was an important aspect of the work of Lunokhod. The vehicle would stop and the penetrating cone would be lowered at the back of the vehicle. The penetrator, called PrOP, coned at an angle of 60°. First, it was forced into the surface to a depth of up to 5 cm to test force. Some pressure was applied, equivalent to a sixth of the weight of the rover. Then vanes inside the cone were rotated 90° for torque, again another measurement of surface strength. This was done 500 times. The results of penetration and trafficability tests were published in detail, finding that Lunokhod operated on surfaces that were much weaker than the lunar roving vehicles driven by the Apollo 15–17 astronauts. In September, penetrometer stops were made every 65 m.

Whatever the power problems might have been, they lifted – for in June Lunokhod headed 1,559 m north–northeast to a further set of four craters, which were set to be its final exploration area. It explored this area thoroughly, and it would seem that Lunokhod's power was really beginning to fail at this time. Total distances travelled were falling: 220 m in July, 215 m in August and only 88 m in September. Earlier, on 5th August, American astronauts David Scott, James Irwin and Al Worden flew directly over the mooncar in their Apollo 15 command module and Lunokhod's magnesium alloy frame glinted in the Sun. As it drove slowly, plodding across the moonscape, the two different moon explorers stood in stark contrast to one other.

Then, suddenly, whilst at work on 4th October exploring a group of four craters far north of the landing site, Lunokhod's 'heart' – its isotope power source – gave out. Telemetry reported a rapid drop in pressure inside the hermetically sealed cabin. The wheels halted, the TV pictures and signals ceased. It was the end.

Considering Lunokhod had been designed to function for only three months and had worked for nearly a year, its mission was a cause of much congratulation. It was the USSR's most brilliant achievement in the field of automatic space exploration. Apparently primitive in design, superbly built, with a reliability the perfectionist buffs in NASA would have envied, it endeared itself to the public at large and became the most exciting robot of its day. In statistical terms alone, its achievement was impressive. It had travelled 10.54 km, covered an area of 80,000 m^2, sent back 20,000 pictures including 200 panoramas and X-rayed the soil at 25 locations. Only later did the Russians reveal that in fact the braking system on Lunokhod had failed quite early in the 'on' position and all the time it was driving against the friction of its brakes. The driving team and the scientists were exhausted. For ten months, they had worked

Table 7.2. The journey of Lunokhod, 1970–1

Start	End	Distance (m)	Notes
17 Nov 1970	22 Nov 1970	197	First journey and charging up of battery
10 Dec 1970	22 Dec 1970	1,522	Journey to the southeast
8 Jan 1971	20 Jan 1971	1,936	Return to landing site
8 Feb 1971	19 Feb 1971	1,573	Heading north
9 Mar 1971	20 Mar 1971	2,004	The longest distance travelled
8 Apr 1971	20 Apr 1971	1,029	Power exhausted from period trapped in crater
7 May 1971	20 May 1971	197	Concentrating on static experiments
5 June 1971	18 June 1971	1,559	Resuming journey north
4 July 1971	17 July 1971	220	Power failing, decision to focus on static experiments
3 Aug 1971	16 Aug 1971	215	
31 Aug 1971	15 Sep 1971	88	
30 Sep 1971	4 Oct 1971		Loss of power, end of mission

ten-hour shifts, 14 lunar days at a time, punctuated by short breaks for the three days of lunar noon when they would enjoy the nearby seaside resort and longer 14-day breaks for the lunar night, when they flew back to Moscow to review their data.

NOW TO THE LUNAR HIGHLANDS

Chief designer Georgi Babakin lived long enough to see the triumph of Lunokhod. He died suddenly in August 1971, aged only 57 and at the very height of his powers. His replacement was N-1 rocket engineer Sergei Kryukov. Sergei Kryukov was born 10th August 1918 in Bakhchisarai in the Crimea, his father being a sailor and his mother a nurse. His mother was ill throughout his early years and died when he was eight. Young Sergei spent much of his childhood in an orphanage, but relatives eventually removed him and ensured he got an education. He caught up quickly, entered Stalingrad Mechanical Institute in 1936 and then its artillery facility, continuing to work there even as the city was under German siege. With the war over, he continued his education in the Moscow Higher Technical Institute while getting work in the # 88 artillery plant there. No sooner had he started than he was transferred to Germany, his task being to reverse-engineer the world's first surface-to-air missile, the *Schmetterling*, which for the Russians was as important as the A-4 surface-to-surface missile. On his return, he transferred to work for Sergei Korolev in OKB-1, where he developed the R-3, R-5 and R-7 rockets, being number four in the design of the R-7 after Korolev, Tikhonravov and Mishin. His contribution was recognized by an Order of Lenin.

His experience with the R-7 was a useful base for working in Lavochkin. After the R-7, Kryukov went on to work on upper stages, principally the Molniya's block I and block L. Assigned to develop the block D for Proton and the N-1, he fell out with Vasily Mishin in 1970, but then managed to transfer to NPO Lavochkin, never imagining that within a year he would become director. He had two deputies: one was responsible for moon probes (Oleg Ivanovsky), while the other was put in charge of planetary probes (Vladimir Perminov).

By the time the Luna 17/Lunokhod mission ended, another Ye-8-5 mission had been dispatched. Luna 18 was launched on 2nd September 1971. After a perfect journey to the moon, it entered a circular lunar orbit of 101 km, inclination 35°, 1 hr 59 min. This was lowered to a pre-descent orbit of 100×18 km and it fired its braking rockets over an area just north of the Sea of Fertility on 11th September. The small thruster rockets tried to guide it into a suitable landing site, but the fuel supplies gave out and it crashed. Not even Radio Moscow felt able or thought it worth its while to invent a cover-up story. Something like 'testing new landing techniques' may have been considered, but this time it admitted that the landing had been 'unlucky' in a 'difficult and rugged' upland area. Although the term 'failure' was not explicitly used, it was one of the few early occasions on which the Russians did not pretend that all mission objectives had been attained. Some scientific data were even obtained from the mission, for scientists were able to infer the density of the lunar soil from the altimeter system and the outcomes were published four years later.

The intentions behind Luna 18 became clear when its backup vehicle was sent aloft on 14th February 1972, entered circular lunar orbit of 100 km on the 18th, 65°, 1 hr 58 min. Luna 20 made a pre-descent orbital firing the following day, bringing it into a path of 100 × 21 km, 1 hr 54 min. The sharper inclination of 65° may have given Luna 20 a safer approach route to the landing site. Luna 20 fired its engines for 267 sec to come in for a landing late on 21st February. This was the critical stage and it had gone wrong twice before. Luna 20's final orbit had a perilune of 21 km. Once this final engine deadstop blast finished, 1.7 km/sec had been cut from velocity and Luna 20 made a rapid descent, coming down at 255 m/sec, much faster than Luna 16.

Luna 20 was coming down right on the top of uplands. The Sea of Fertility lies on the right of the moon's visible face and Luna 16 had landed on one of its flattest parts. To the north, hills rise and there are soon mountains 1,500 m high. It was in a small plateau between two peaks where Luna 20 was aimed, less than 1,800 m from where its predecessor had come to grief on a sharp slope. The area is called Apollonius. It was tougher than anything the American Lunar Module would have tried. Because of the much higher descent rate, the propulsion system fired sooner – at 760 m – and Luna 20 made it, whether through luck or skill we do not know.

And so it came to rest, straddled by towering mountain peaks. Signals at once indicated to relieved controllers that it was safe and secure. Within seven hours, aided by a small television camera, its drill was hard at work scooping up lunar soil. Unlike Luna 16, Luna 20 landed in daylight and a picture of the drilling was subsequently published in the Soviet press [8]. Two cameras were installed on the landing stage, with a viewing angle of 30°. The drill rotated at an anti-clockwise 500 r.p.m., cutting away with sharp teeth which put material into a holding tube. It had two engines: one for the main drilling, but a second to take over if it faltered. The drill was kept sealed until the moment of drilling began, for it was important to keep it lubricated right up to the moment of operation. If it were exposed to a vacuum too early, there was the danger that the lubricant would evaporate.

Luna 20 view of surface

The drilling operation took 40 min and was photographed throughout. The rig encountered stiff resistance at 10 cm and operations had to stop three times, lest it overheat. When it reached 25 cm, the samples were scooped into the return capsule to await the long journey home. The retrieval took 2 hr 40 min in the end and was probably the most difficult of all the sample recovery missions. The conditions were undoubtedly tough and the sample probably much smaller than hoped for.

The cameras swivelled around to take an image of the surrounding moonscape, with Earth rising in the distance. The onboard computer fired the engines early on 23rd February and the return vehicle climbed away from the lunar peaks. Once again, the Kazakhstan landing site required a lunar liftoff when the moon was over the Atlantic. So, 2.84 days later it headed into reentry, the small cabin separating 52,000 km out. Amateur trackers picked up signals from Luna 20 growing in strength as it approached the Earth. Both the ascent spacecraft and the cabin came in quite close to one another, signals fading out only 12 min before touchdown [9].

Despite a steep reentry angle of 60°, twice that of Luna 16, only 5 mm of ablative material burned away. An appalling blizzard hit the recovery area that day. Helicopters spotted the tiny capsule – parachute, antennae and beacon deployed – heading straight into the Karakingir River some 40 km northwest of Dzhezhkazgan at 48°N, 67.6°E. Would the precious samples be lost at this stage? Luckily, the capsule came to rest on an island in the middle of the river and in a snowdrift and trees. But getting it back was easier said than done. The gale was too severe for the helicopters to land. Four cross-country vehicles tried to get across on the ice but it cracked so they called it off for fear of falling in. Their crews eventually retrieved the battered and burnt capsule the next day when the wind abated. Its contents were opened at the Academy of Sciences. They were surprisingly small – between 30 and 50 g. But it was moondust all the same and the light ash-gray dust was 3bn years old. The records state it consisted mainly of anorthosite, with olivine, pyroxene and ilmenite. High-quality non-rusting iron was found, one of the most interesting findings. The colour was lighter and had more particles than the previous sample. Luna 20's samples had the highest content of aluminium and calcium oxides of all the moon samples. Two grams of Luna 20 samples were exchanged with American Apollo 15 samples. The Americans were able to provide accurate dating of the Soviet sample. Seventy chemical elements were found, with an average density of $1.15 \, g/cm^3$.

ALONG THE RILLE OF LE MONNIER BAY

Apollo ended in December 1972, and from thereon the Russians knew that they had the moon to themselves. When Luna 21 headed moonwards on 8th January 1973, the launching was seen in the West as deliberately calculated to take advantage of the end of the Apollo programme. In fact, the timing was coincidental. The second mooncar, for that was what Luna 21 carried, had taken a full year to redesign after Lunokhod had terminated its programme. Luna 21 weighed 1,814 kg and its translunar flight was problematical. False telemetry signals nearly aborted the mission and then Lunokhod 2's solar lid opened during the translunar coast, without being asked to do so.

Luna 20 landed in snow

Lunokhod 2

Luna 21 entered a near-circular lunar orbit on schedule on 12th January between 90 km and 110 km, 1 hr 58 min, 60°. The next day, the perilune was lowered to 16 km. On its 41st moon orbit, 255 km from its objective, Luna 21 began its descent from an altitude of 16 km, coming down at 215 m/sec. The target was the 55 km wide Le Monnier cratered bay, the target for the first, failed Lunokhod in 1969. Le Monnier was only 180 km from the valley just visited by Jack Schmitt and Eugene Cernan of Apollo 17. Off the edge of the Sea of Serenity, the now eroded remains of the Le Monnier crater cut into the edge of the rocky Taurus Mountains. The main engine blasted at 750 m, cutting out at 22 m when a secondary thruster brought the spacecraft down to 1.5 m, from which height it fell gently to the surface at 7 km/hr.

Luna 21 came down in a relatively flat area surrounded by the high rims of the old crater. The site had been chosen because Le Monnier marked the transition between the low *mare* and the upland continental area. Le Monnier was a flooded rim rather than a sharply defined crater. The location was 25.85°N, 30.45°E, the landing time 02:35 Moscow time on 16th January. The navigation system failed at the moment of touchdown, which meant that – although the rover was intact – the drivers were not sure exactly where it was.

Lunokhod 2 first activated its cameras and panned around the landing site from the high vantage point of the landing state. With the slogan *Fifty years of the Soviet*

Lunokhod 2 with hills behind

Union! 1923–1973 emblazoned on it, Lunokhod 2 rolled down the landing ramps not long afterwards. Lunokhod 2 at once made a trial journey over the surface and then parked for two days 30 m away to charge up the batteries. Cameras at once showed the *mare*, the crater rim to the south and a massive stone split into lumps in the foreground.

Lunokhod 2 was a distinct improvement over its predecessor. It was 100 kg heavier at 840 kg. It could travel at twice the speed, having two speeds, 1 km/hr and 2 km/hr (its average turned out to be 15.5 m/hr). Lunokhod 2 was designed to handle obstacles of 40 cm and holes of 60 cm. It had twice the range. Addressing some of the driving problems arising from the first Lunokhod, pictures were now transmitted to its drivers every 3.2 sec, compared with 20 sec before. The cameras were moved much higher. Lunokhod 2 had three low-rate cameras on the front, able to scan 360° vertically and 30° horizontally, with two double panoramic cameras able to scan 180° horizontally and 30° vertically. The television cameras had three possible scan rates: 3.2, 5.7 and 21.1 sec per frame. There were new scientific instruments, most notably a photodetector called *Rubin* to detect ultraviolet light sources in our galaxy and the level of Earthglow on the nighttime moon. The heat source was again an isotope made of polonium-210. The magnetometer was deployed on a boom 2.5 m in front. The laser reflector had an accuracy of 25 cm.

Lunokhod 2's programme was first to inspect the descent stage, to which it would not return and then it would head south to the mountains 7 km away and explore there. In its first journeys, Lunokhod 2 investigated craters close to the landing stage, taking detours to avoid big rocks. At the end of its first lunar day, Lunokhod 2 parked 1 km southeast from its landing stage.

Lunokhod 2 instruments
- Soil mechanics tester.
- Solar X-ray detector.
- Magnetometer.
- Photodetector (*Rubin*).
- Laser reflector (France).

Lunokhod 2 wheel marks on the moon

On 8th February, Lunokhod 2 began its first full lunar day and in ten days reached the southern rim of Le Monnier, exploring the edge of the rim and two craters there, one 2 km wide. The nature of the ground varied from soft and loose to hard and firm. In one crater in the foothills, it circled around the edge of a crater with a 25° slope, taking analyses at numerous spots. In the course of one of its early sessions, the bug-eyed roving vehicle went 1,148 m in six hours – much faster than anything achieved before. It climbed one hill of 400 m, with its wheels at one stage slipping up to 80%. From the top it sent back an eerie photograph of the Taurus peaks glowing to the north, 60 km away and the thin sickle of the Earth rising just above. As it journeyed, it measured and analyzed the lunar soil. Lunokhod 2 rambled around the southern rim of Le Monnier. By the end of February, the rover had travelled farther than the first Lunokhod in its ten months.

Now, in March, Lunokhod 2 headed off on its greatest adventure. On the day the Lunokhod landed, the Institute of Space Research (IKI) in Moscow was holding a symposium on solar system exploration, one also attended by American scientists. One of the Americans had brought with with him a batch of new Apollo 17 pictures of that region of the moon, for it was close to the Apollo 17 landing site, giving them to Lavochkin lunar bureau chief Oleg Ivanovsky. They were so detailed that they showed up a new rille, 16 km long and 300 m wide, the Fossa Recta, to the east of the rover. Lunokhod 2 was duly directed there, setting out on an eastward course to explore the new rille and there it travelled, the flat crater of Le Monnier on its left flank, its rim and the Taurus Mountains on its right. The rover skirted around small craters as it journeyed eastward. By 20th March it was just west of the rille.

On lunar day 4, April, Lunokhod 2 travelled southwest around the southern end of the rille, exploring it from both sides. Here, geologists were excited to spot an outcrop of bedrock. Although the surface was a firm volcanic basalt, there were occasional dusty soft spots and at one stage the tracks of the rover sank about 20 cm into the lunar surface. The journey along the rille was a dangerous one, for there were many metre-size boulders along its ledge. Lunokhod parked around the eastern side on 19th April. The magnetometer identified a magnetic anomaly on the western edge of the rille.

The ground control veterans of Lunokhod, temporarily idle for a year, had resumed their work in two shifts of two hours each. Driving the Lunokhod over the moon required teamwork. The navigator was responsible for the route over the

Driving the Lunokhod

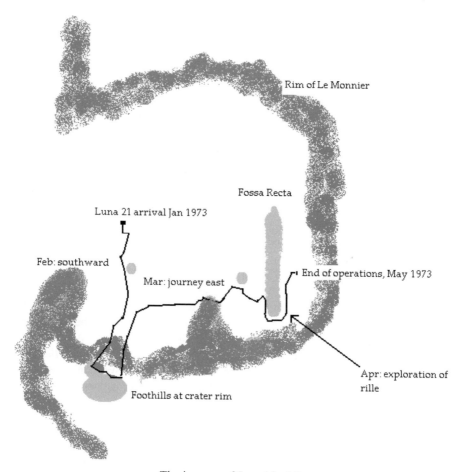

Rim of Le Monnier

Fossa Recta

Luna 21 arrival Jan 1973

Feb: southward

Mar: journey east

End of operations, May 1973

Apr: exploration of rille

Foothills at crater rim

The journey of Lunokhod 2

lunar surface, but in reality the decisions were taken by the whole team, once even voting on the best route. The radio and antenna operator was responsible for ensuring that – whatever the direction of the rover – the antenna was pointing toward Earth and the solar lid was in a position to collect sufficient energy from the sun. The teams worked their shifts according to the lunar day and night and sometimes lost track of Earth time, day and night. Once, during an especially tense drive across the moon, contact with Lunokhod broke off abruptly. Only when they pulled back the curtains of the control room did that realize that the reason was that the moon had just set below the horizon.

In May, Lunokhod resumed its journey on the eastern side of the rille, once traversing up an 18° slope. It was expected that Lunokhod 2 would continue its work for several months, but on 4th June came the sudden announcement from Radio Moscow that 'the programme had been completed.' No explanation was given at the time, nor a suggestion that anything had gone amiss. Like so many events in the Soviet lunar programme, the real story did not emerge for ages – 30 years later [10]. Some extreme versions even came out, like that the moon rover had 'turned turtle'.

The reality was more prosaic. On 9th May, soon into the new lunar day, Lunokhod had descended into a small but deep 5 m wide crater inside a much larger crater, its depth concealed by the shadows thrown by a low-sun angle. As the drivers tried to manoeuvre Lunokhod out of the crater, the lid touched the crater wall, dumping clumps of lunar soil onto the solar cells. The immediate consequences were not serious – the loss of some electric power – but the long-term consequences were fatal. When night came, the lid was closed, as normal. In closing the lid, the soil was then dumped onto the radiator. When lunar dawn came in early June and the lid was raised again, the lunar soil acted as an insulator, preventing the rover from properly releasing its heat. The heat inside rose and Lunokhod quickly died.

The Russians seem to have been disappointed, but there is little reason why they should have been. Lunokhod 2 had travelled 37 km, sent back 86 panoramic pictures and 80,000 television pictures and had covered four times the area of its predecessor. It had investigated not only crater floors but much more difficult geological features like rilles and uplands. One of its most interesting findings actually had nothing to do with the lunar surface, but the suitability of the moon as a base for observing the sky. Whilst it would be excellent during the lunar night, during the daytime the lunar sky was surrounded by a swarm of dust particles, a kind of atmosphere that would make telescopic observations very difficult. The astrophotometer determined that the lunar night sky was, in Earthglow, 15 times brighter than Earth's night sky in moonlight. Detailed tables of the composition of the lunar rocks were published, comparing those sampled by Luna 16, Lunokhod 1 and Lunokhod 2. In the case of Lunokhod 2, it was possible to make comparisons between the composition of the *mare* and continental rocks [11], the proportions of aluminium, silicon, potassium and iron being different. The RIFMA-M X-ray flourescent spectrometer measured the soil near the lander as 24% silicon, 8% calcium, 6% iron and 9% aluminium, but the soil at the edge of the Taurus Hills showed a sharp rise in iron content. Lunokhod 2's average speed was 27.9 m/hr, seven times faster than its predecessor and it covered seven times the distance. Up to a thousand penetrations of the lunar surface were made by the two rovers between them. The laser fired 4,000 times with an accuracy of 25 cm. The magnetometer had been installed following measurements of some form of lunar ionosphere by Luna 19 the previous year. Lunokhod 2's magnetometer determined

Table 7.3. The journey of Lunokhod 2, 1973

Start	End	Distance (m)	
16 Jan	24 Jan	1,260	Land and charge battery
8 Feb	23 Feb	9,806	Journey south to the rim of Le Monnier Bay
11 Mar	23 Mar	16,533	Journey east to Fossa Recta
9 Apr	22 Apr	8,600	Exploration of the rille
8 May		800	End of mission

that there was a very weak permanent magnetic field around the moon, its measurements being broadly in line with those of Apollo. The temperature of the lunar night was measured at $-183°$C. Back in Moscow, mission scientists made a geological map of Le Monnier Bay, complete with slices of the surface, bedrock and underlying strata.

MOON ORBITERS

The Ye-8 series included two orbiters, Ye-8LS, both being launched successfully. They flew the last of the trio of rovers of orbiters and rovers, although it had originally been intended they go first. Their role was to:

- Take photographs of points of interest so as to identify landing sites for later sample return, rover and manned missions.
- Study mascons, magnetic fields, the composition of lunar rocks, meteorites and cislunar space.

New cameras were developed for the series by Arnold Selivanov. Essentially, he adapted the optical–mechanical camera of Luna 9 and 13 as an orbital panoramic camera in such a way as to make 180° long panoramic sweeps extending to the edge of the moon. The images would be developed on board, scanned at 4 lines/sec and relayed back to Earth. These are called optical–mechanical linear cameras and can be used from moving spacecraft.

Warning of a new moon probe first appeared in January 1971 when predictions of 'low-flying artificial satellites' were made that would fly 'fairly soon'. Sure enough, Luna 19 was launched on 28th September 1971 and entered circular lunar orbit of 140 km at 40°, 2 hr 01 min, on 3rd October. Two sets of details were published for the

Ye-8LS

first day of operation, indicating either a tweaking of the orbit or a refinement of the earlier figures. Three days later, it settled into steady operational orbit of 127×135 km, 2 hr 01 min, $40°$. It is more than likely that Luna 19 kept the large tanks used for orbital insertion and continued to use them for manoeuvres, rather than drop them soon after arrival in lunar orbit. The mission was publicized through periodic reports in *Pravda* and *Izvestia*. Although at least five full panoramas were assembled, only one section of one was published, along with an illustration showing the probe being loaded onto its Proton carrier rocket, but the detail is poor.

The mission lasted till 3rd October 1972 and 1,000 communication sessions were held. Luna 19 reported back on magnetic fields, mascons, the lunar gravity field, meteoroids and sent back televised pictures of an area $30°$S to $60°$S and $20°$E to $30°$E, the quality of publication much improved compared with Luna 12 in 1966. In February 1972, it swept over the Torrid Gulf near the crater Eratosthenes ($11°$W, $15°$N) and filmed rock-strewn plains above which reared a volcanic-like summit. In order to take such pictures it had dropped into a new, lower orbit of 77×385 km, 131 min. Another landing area surveyed was around craters Godin and Agrippa at $10°$E, $3°$N. Some science reports were issued, noting how Luna 19 had measured solar flares and plasma, mascons, the lunar surface and the composition of its soil. The strength of the magnetic field on the nearside and farside of the moon was compared. Radiation levels were measured, especially their rise and fall during solar flares. Ten solar flares were detected. Some cislunar plasma was detected, but the outcome of this experiment was unclear. An altimeter called Vega was carried to measure the precise distance of the probe to the moon (important during its low perilunes). A gamma ray spectrometer took broad measurements of the composition of the lunar surface. A radio occultation experiment was carried out in May 1972 and this found charged particles about 10 km over the moon. The magnetometer measured magnetic fields as the moon moved in and out of the Earth's long magnetic tail. The mission lasted 4,000 orbits.

Luna 19 low pass

It was a full year before the next orbiting moon probe, Luna 22, took off on 2nd June 1974. The Luna 22 launch came at an important international moment, for the first Soviet–American conference on lunar exploration took place that month, June 1974. Together, the scientists were able to agree on the approximate date of the moon (4bn years), the nature of its crust (thick), the processes that had shaped it and that the moon shared a broadly similar formation to the Earth.

Ground observatories tracked Luna 22 as far as 250,000 km out. Luna 22 entered almost circular moon orbit at 219×221 km, 2 hr 10 min, 19.6° four days later. A week later, it swooped down to 25×244 km for special photography for four days, before going back up again to 181×299 km. Over the next year, Luna 22 several times altered its orbit, displaying both versatility and reliability. In November 1974, coinciding with the arrival in orbit of Luna 23, it operated in an eccentric orbit of $171 \times 1,437$ km out, 3 hr 12 min, then raising its perilune to 200 km and making a minor plane change to 21°. Then, in August 1975 it dipped to a mere 30 km over the surface for a week, going out farther to 1,578 km, before returning to a regular orbit of $96 \times 1,286$ km out when its mission ended in November 1975.

Lunar orbit photography was done both from altitude and at low points, the latter presumably to search for landing sites, but no details were ever given of the sites surveyed and the following two Lunas (23 and 24) were both aimed at the Luna 15 sites which, presumably, had been mapped before 1969. There were two extended periods in which no manoeuvres were made, presumably so as to give time to measure changes to its path arising from distortions in the moon's gravitational field.

Few scientific results were released from the mission, although they could have been substantial, as evidenced by the heavy radio traffic to and from the probe over the 18 months of its operation. These results could have covered the surface composition, topography and micometeoroid impacts, which were much fewer in the higher orbit. Lunar topography was mapped carefully through the use of an altimeter and a gamma ray spectrometer analyzed the composition of the surface [12]. Science reports indicated that Luna 22 studied the moon's gravitational field, micrometeorites (23 impacts recorded) and solar plasma. The probe indicated that a sheath of ionized gas forms 8 km over the lunar surface during sunlight. Eight photographs eventually reached the NASA archives in the 1990s. Ten full panoramas were reportedly assembled.

Orbits of Luna 19, 22
Luna 19

2 Oct 1971 (LOI)	140×148 km, 2 hr 04 min, 40.58°
2 Oct 1971	140 km circular, 2 hr 01 min, 40.58°
7 Oct 1971	127×135 km
28 Nov 1971	77×385 km, 2 hr 11 min
2 Dec 1971	127×135 km
Feb 1972	77×385 km, 2 hr 11 min

Luna 22

6 June 1974 (LOI)	219 × 221 km, 2 hr 10 min, 19°
9 June 1974	25 × 244 km for four days
13 June	181 × 299 km
11 Nov 1974	171 × 1,437 km, 3 hr 12 min, 19.55°
2 Apr 1975	200 × 1,409 km, 3 hr 12 min, plane change to 21°
24 Aug 1975	30 × 1,578 km
2 Sep 1975	Orbit raised to 96 × 1,286 km, 21°

The orbital paths of the two missions show similarities and differences. Having adjusted its original insertion orbit, Luna 19 operated for the first portion of its mission from a 127 × 135-km near-circular orbit (October–November). At the end of November, it dropped its perilune to 77 km for three days of photographic observations, before coming back to the circular orbit. In February, Luna 19 went back to its lower perilune, where it apparently stayed. Luna 22, by contrast, followed three sets of orbits. Its operating orbit was around 200 km, dropping twice for photographic surveys for periods of less than a week, in late June 1974 and late August 1975. The perilunes were on both occasions much lower than those of Luna 19, this time descending to 25–30 km. In addition, Luna 22 also flew, twice, into an eccentric orbit, out as far as 1,578 km. The precise rationale for these manoeuvres has never been explained.

RETURN TO THE SEA OF CRISES

For the rest of the series, the Ye-8-5 was redesigned as the Ye-8-5M. The chief improvement was a much more versatile rail-mounted drill for obtaining samples. This drill was a radical improvement on its predecessors which could only reach 30 cm and the new one was able to penetrate to a depth of no less than 2.5 m. This assignment went to the General Construction Design Bureau of Vladimir Barmin (1909–1993). Barmin was a close colleague of Sergei Korolev and a member of the original council

Vladimir Barmin

of designers of 1946. He was the constructor of the cosmodromes, a task of enormous proportions involving the heaviest Earth-moving and digging machinery in the world. Now he got the assignment to make precision drilling equipment for use on another world.

It is possible that the Ye-8-5M missions benefited from the studies of the lunar gravitational environment by Luna 19 and 22. This time the target was the old Luna 15 site at 13°N, 62°E in the large Sea of Crises, a region never explored by the Americans.

Three Ye-8-5Ms were launched, in October 1974, October 1975 and August 1976. Luna 23 entered a lunar orbit of 94×104 km, 1 hr 57 min, 138° (12° more than Luna 15) on 2nd November 1974, adjusted on the 6th to a pre-descent orbit of 17×105 km. When it tried to land in the southern part of the Sea of Crises on its 50th revolution, it was severely damaged in the course of the landing. The soil-collecting gear was wrecked, although the descent stage was able to continue transmissions for a further three days and contact was lost on the 9th November. It was normal for the descent craft to continue to transmit on 922 MHz for this period, though for what purpose is uncertain, except to relay radiation measurements back to Earth.

It may or may not have been open to the Russians to send the empty return craft back to Earth anyway, but the manoeuvre was not attempted. A replacement mission was organized, but the next Luna failed a year later due to block D failing to ignite.

Finally, Luna 24 entered a circular orbit of 115 km, 1 hr 59 min, 120° on 14th August, adjusted to a pre-descent orbit of 120×12 km on the 17th, the lowest of any pre-landing orbits. Amateur trackers in Sweden and Florida picked up its signals on 922 MHz for 20 min on every orbit as it transmitted back to Earth. Luna 24 came down in darkness close to the wreckage of Luna 23 and, it is suspected, at the exact place of Luna 15's targeted spot, 17 km from the small crater Fahrenheit. Touchdown was on 18th August 1976 and all went well this time. As the rotary percussion rig drilled into the soil, the sample was stored in a rubber pipe in such a way as to prevent clogging and compression. The drill brought up samples weighing 170 g in a 2.6 m long core sample and had been modified in order to minimize grains falling off.

Back on Earth, the same amateur trackers were listening in to Luna 24's liftoff [13]. Normally, the ascent rocket would begin transmission on 183.6 MHz from the moment of engine burn and continue to transmit during the ascent from the moon. Luna 16 and 20 had spent 1.1 and 1.15 days on the moon respectively, so the same could be expected of Luna 24. But this is not what happened. Instead, Luna 24 lifted off early, after only 0.95 days. For the first time, the lunar liftoff took place with the spaceship in line of sight of Yevpatoria at the moment of liftoff. Hitherto, these liftoffs had taken place when the moon rocket was not in sight of Yevpatoria (though it could be seen by an Atlantic tracking ship), but the final stage of the return journey was in line of sight, which was more important. This time, the Russians must have felt so confident with the return trajectory that it could be accomplished out of sight of Yevpatoria. The return flight was longer than the previous missions, 3.52 days and the spacecraft came back into the atmosphere in a curving trajectory around the back side of the Earth like Zonds 6 and 7, with a recovery zone in Siberia, one never used

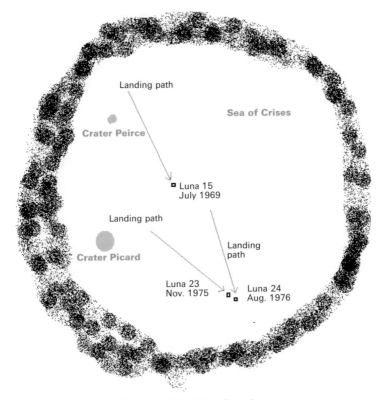

Luna 15, 23, 24 landing sites

before or since. The capsule came down in summertime Siberia 200 km southeast of the tundra town of Surgut and no difficulty was reported in finding it.

Samples were again exchanged with the Americans (3 g) and they were dated to 3.3bn years. Some samples also went to Britain. The post-mission report, given in *Pravda* on 5th September, related how 60 different chemical elements had been found, dark grey to brown in colour. They appear to be laid down in layers.

Outcome of sample return missions

Date	Spacecraft	Landing site	Samples (g)	Type
Sep 1970	Luna 16	Sea of Fertility	105	*Mare*
Feb 1972	Luna 20	Apollonius	50	Upland
Aug 1976	Luna 24	Sea of Crises	170	*Mare* core sample
			325	

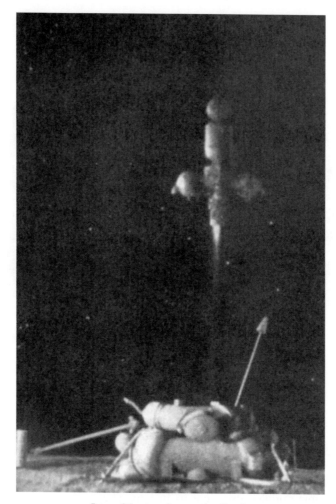

Luna 24 returning to the Earth

Ye-8 series: scientific outcomes

Characterization of lunar soil from three locations: *mare, mare* core sample and uplands.

Characterization, penetration, measurement of lunar soil *in situ* from two *mare* locations (Lunokhod, Lunokhod 2), studying density, strength, composition.

Refinement of lunar and interplanetary gravitational field.

Fluxes in radiation levels on moon and in moon orbit over a period of months.

Measurement of precise distances between Earth and the moon.

Characterization of local lunar environment in Bay of Rains, Le Monnier.

Mapping of selected areas on lunar nearside.

Measurement of dust levels over daytime lunar surface.

Discovery of thin sheath of ionized gas over sunlight side of lunar surface.

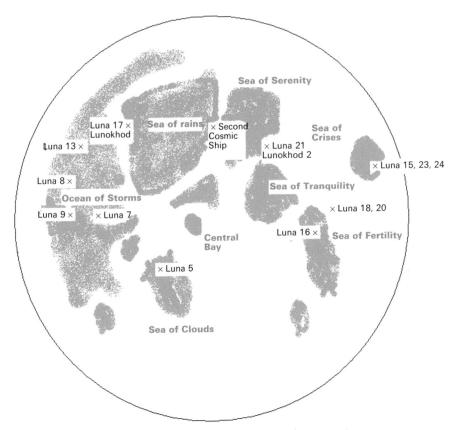

Location of Soviet moon probes

WINDING DOWN THE PROGRAMME OF AUTOMATIC LUNAR EXPLORATION

Luna 24 was the last moon mission by the Soviet Union or Russia. Its return cabin, along with those from Luna 16 and 20, was given to the Lavochkin Museum. Twenty-five such cabins had been built altogether, both for flight and tests. Three still rest on the moon (Luna 15, 18, 23) and five were lost in rocket explosions.

During the period September 1970 to June 1973, a series of missions was promised to build on the successes of the lunar sample return and the Lunokhod. On Luna 16's return, the Soviet media announced that the 1970s would be 'the decade of the space robot'. Among the missions spoken about were:

- Sample return missions from remote areas, including uplands and the poles.
- Lunokhods to carry drills to obtain cores and analyze them in onboard laboratories.
- Lunokhods to collect rocks and deliver them to sample return missions.

- Telescopes on the lunar farside.
- Automatic static observatories on the moon.
- Use of relay satellites to control and receive information from farside moon probes.

The use of Lunokhods to deliver rocks to a sample return craft would have been impressive. For this, the Lunokhod would have been fitted with a robot arm. Such a mission was sketched in detail and the rover would have been called *Sparka*, Russian for 'pair'. Further into the future, VNII Transmash envisaged a 'heavy Lunokhod' which looked like an armoured personnel carrier, 4.7 m long, 4.3 m wide, with eight wheels, panels and dish aerials on top, eight 1,200 mm wheels and able to traverse very extensive distances [14].

The extensive discussion of plans for future moon probes in the Soviet press came to an abrupt end in June 1973. References to rovers were now made in the context of their achievements being used to design Mars rovers, rather than future moon rovers. The Soviet media barely reported the last set of missions. It seems that the decision was taken in summer 1973 to wind down the Ye-8 moon programme over the next four years, using up most of the already built hardware. Lunokhod 3 was built and ready to fly in 1977 as Luna 25, but ended up instead in the Lavochkin Museum. Lunokhod 3 was similar to Lunokhod 2, but with an improved camera system. The Proton rocket that should have brought it to the moon was given over to a communications satellite instead.

When Luna 24 returned to Earth, there was no official indication that the programme of unmanned lunar exploration had drawn to a close and, of course, the cancellation at the same time of the N1-L3M programme was not announced either. One winner from Luna 24 was Vladimir Barmin, who was now charged with developing a drill to dig into the rocks of Venus, his new machinery being carried on the forthcoming Venera 11 and 12 missions in 1978.

Lunokhods roam the moon

Lunokhod 3

When Lunokhod 3 was cancelled, the lunar team was dispersed to the Venus missions. Oleg Ivanovsky, the deputy director and responsible for lunar probes, was put in charge of building an orbiting astronomical observatory, called Astron. Once this flew, successfully, in 1983, he retired, taking up a new voluntary post as head of the Lavochkin Museum. Other staff were assigned to other probes and missions.

All the scientists could do was content themselves with publishing the results of their investigations, both in the Soviet press and in collaborative publications abroad.

Valeri Barsukov

Despite the heat of the moon race, scientists from the two countries were eager to share and compare the results of their analysis of the results of the moon missions. Many geologists made extensive cross-comparisons of the differences between the three Luna samples and the six Apollo samples, classifying them according to origin, type and composition. Even though the Luna samples were small, they were three distinct types: *mare*, highland and core. The glassy features of the Luna 16 rock were especially unusual. In the year after Luna 24's return, NASA published the proceedings of the Soviet–American conference on the geochemistry of the moon and planets [15] and Soviet papers were published in other Western outlets, such as the journal *The Moon*. The NASA papers included the analysis of the moonrock collected by the Lunas and various articles, ranging from studies of the rocks from an individual mission to broader reviews, such as T.V. Malysheva's *The problem of the origin of the lunar maria and continents*. Lunokhod 2 produced a rich seam of scientific papers, such as L.L. Vanyan's *Deep electronic sounding of the moon with Lunokhod 2, Measurement of sky brightness from Lunokhod 2* and Dolgov *et al.*'s: *The magnetic field in Le Monnier Bay according to Lunokhod 2*. Kiril Florensky's *Role of exogenic factors in the formation of the lunar surface* included a series of hitherto unseen Lunokhod 2 pictures. The results of the very last mission were published by Nauka as *Lunar soil from the Mare Crisium*, by Valeri Barsukov, in 1980.

As for Alexander Kemurdzhian, the designer of the moonrovers, he wrote another thesis about his creations, obtaining a second doctorate and the title of professor. His STR-1 robot was involved in the investigation and cleanup of the Chernobyl nuclear disaster. Kemurdzhian exposed himself to so much radiation there that he had to be treated in the Moscow #20 hospital afterwards. He wrote 200 scientific works and patented 50 inventions. Almost eighty, he retired in 1998, though colleagues noticed little change in his output or energy and he was the chief speaker at the 30th anniversary of the Lunokhod meeting held in Tovstonogov in November 2000. His health deteriorated soon after this and he died on 24th February 2003 in the hospital which had treated him for radiation burns. Alexander Kemurdzhian was buried in the Armenian part of the Smolensky Cemetery in St Petersburg. Asteroid

#5993 was named after him, and the International Biographic Centre named him one of the outstanding people of the 20th century.

Final round of moon missions
Sample return Ye-8-5 missions
14 Jun 1969	Failure
13 Jul 1969	Luna 15 (failure)
23 Sep 1969	Failure (Cosmos 300)
22 Oct 1969	Failure (Cosmos 305)
19 Feb 1970	Failure
12 Sep 1970	Luna 16
2 Sep 1971	Luna 18 (failure)
14 Feb 1972	Luna 20

Sample return Ye-8-5M series
28 Oct 1974	Luna 23 (failure)
16 Oct 1975	Failure
9 Aug 1976	Luna 24

Lunokhod (Ye-8) missions
19 Feb 1969	Failure
10 Nov 1970	Luna 17/Lunokhod
8 Jan 1973	Luna 21/Lunokhod 2

Orbiting (Ye-8LS) missions
28 Sep 1971	Luna 19
2 Jun 1974	Luna 22

The Ye-8 series did eventually provide the Soviet Union with some form of credible alternative to Apollo and saved some face. The two Lunokhods attracted the most public attention and probably made the most popular impact. They were sophisticated vehicles of exploration and it was a loss to science that Lunokhod 3 was not flown. The soil sample return mission series, although technically difficult and impressive in their own right, cannot be said to have been a great success and the gains were achieved for a disproportionate effort. Although three missions did bring lunar samples back, their haul was small at 325 g, compared with Apollo's 380 kg, while seven missions had failed altogether. The Ye-8LS lunar orbiters may well have achieved solid results, but they were poorly publicized or disseminated. The heart seems to have gone out of the programme in June 1973 and one has the impression that permission was given to fly already-built hardware on the understanding that there would be no further missions thereafter for the foreseeable future. By the time Luna 22 flew, the N-1 programme had been suspended and there was little reason to draw attention to the lunar programme generally. It is probably no coincidence that the last mission, in August 1976, took place only months after the N-1 was finally cancelled in March 1976. It seems that both the manned and unmanned programmes were run down in parallel.

ACHIEVEMENTS OF THE SOVIET AUTOMATIC PROGRAMME FOR LUNAR EXPLORATION

Although the Soviet Union lost the race to the moon, a considerable body of knowledge of the moon was accumulated by the Soviet space probes that flew there over the years 1959–1976. The following is a broad outline of the type of scientific information collected:

- Maps of the farside and limbs of the moon were compiled on the basis of photographs taken by the Automatic Interplanetary Station, Zond 3 and the Zond 5–8 missions. Mapping of selected areas of the nearside was carried out by Luna 12, 19 and 22.
- The environment of near-moon space was characterized by the orbiting missions: Luna 10–12, 14, 19 and 22. Data were obtained on the levels of solar and cosmic radiation, cosmic and solar particles and gravitational fields. The moon's surface was studied from orbit by instruments on Luna 10, 11, 12 and 18.
- The moon's gravitational field was first mapped by Luna 10, then in detail by Luna 14 and refined by Luna 19 and 22.
- Attempts to identify and then measure the moon's magnetic field were made by the First and Second Cosmic Ships.
- The chemical characteristics, composition and density of moonrock were determined *in situ* by Lunokhod and Lunokhod 2 and through samples brought back to Earth by Lunas 16, 20 and 24. The lunar samples, although small, were shared internationally.
- The physical properties of the surface were determined by Luna 13 and the two Lunokhods (RIFMA).
- Precise distances between Earth and the moon were measured by laser reflectors on Lunokhods 1 and 2.
- Radiation levels and temperatures on the surface of the moon were measured by Luna 9, 16, 20, 23 and 24.
- The nature of the lunar micro-atmosphere was measured by Lunokhod.
- The effects of Earth–moon space on animals and other biological samples, especially in respect of radiation, were measured by Zond 5 and 8. These remain the only (non-human) lunar biology missions.

As a result of the American and Soviet efforts, the moon became, unsurprisingly, the best known body in the solar system after our own Earth. With such an improved level of knowledge, the scientific case for returning to the moon became more and more difficult to make, granted our much less comprehensive knowledge of the inner planets Mercury, Venus and Mars and the outer planets. In fact, the moon still had many surprises in store, as the American Lunar Prospector was to prove as it began to search for ice. Scientific instruments by the new century were now able to return much more sophisticated data and at a much higher rate than had been possible in the 1960s, opening up further possibilities for scientific exploration. Miniaturization made it possible for smaller spacecraft to be launched on much less expensive rockets.

The last achievement – Luna 24 core sample

REFERENCES

[1] Clark, Phillip S.: Masses of Soviet Luna spacecraft. *Space Chronicle, Journal of the British Interplanetary Society*, vol. 58, supplement 2, 2005.

[2] Gatland, Kenneth: *Robot explorers*. Blandford, London, 1972;
Ulivi, Paolo: *Moon exploration – an engineering history*. Springer-Verlag, London, 2003.

[3] Ball, Andrew: *Automatic Interplanetary Stations*. Paper given to BIS, 7th June 2003.

[4] Chaikin, Andrew: The other Moon landings. *Air and Space*, vol. 18, #6, February/March 2004.

[5] Carrier, W. David III: *Soviet rover systems*. Paper presented at Space programmes and technology conference, American Institute of Aeronautics and Astronautics, Huntsville, AL, 24th–26th March 1992. Lunar Geotechnical Institute, Lakeland, FL.

[6] Cirou, Alan: L'histoire secrète des Lunokhod. *Ciel et Espace*, septembre 2004 (avec Jean-René Germain).

[7] Tyulin, Georgi: Memoirs, in John Rhea (ed.): *Roads to space – an oral history of the Soviet space programme*. McGraw-Hill, London, 1995.

[8] Vasilyev, V.: Drilling in the lunar highlands. *Nedelya*, 21st–27th February 1972.

[9] Grahn, Sven: Radio systems used by the Luna 15–24 series of spacecraft, *http://www.users.wineasy.se/svengrahn/histind*, 2001.

[10] Chaikin, Andrew: The other Moon landings. *Air and Space*, vol. 18, #6, February/March 2004.

[11] Surkov, Yuri: *Exploration of terrestrial planets from spacecraft – instrumentation, investigation, interpretation*, 2nd edition. Wiley/Praxis, Chichester, UK, 1997.

[12] Huntress, W.T., Moroz, V.I. and Shevalev, I.L.: Lunar and robotic exploration missions in the 20th century. *Space Science Review*, vol. 107, 2003.

[13] Grahn, Sven: *Tracking Luna 24 from Florida and Sweden, http://www.users.wineasy.se/svengrahn/histind*, 2001.

[14] Kemurdzhian, A.L., Gromov, V.V., Kazhakalo, I.F., Kozlov, G.V., Komissarov, V.I., Korepanov, G.N., Martinov, B.N., Malenkov, V.I., Mityskevich, K.V., Mishkinyuk *et al.*: Soviet developments of planet rovers 1964–1990. CNES and Editions Cepadues: *Missions, technologies and design of planetary mobile vehicles*, 1993, Proceedings of conference, Toulouse, France, September 1992.

[15] Pomeroy, John H. (ed.): *Soviet–American conference on the geochemistry of the moon and planets*. NASA, Washington DC, 1977, in two parts.

8

Return to the moon

The cancellation of Luna 25 in 1977 marked the end of the Russian programme of lunar exploration. Nevertheless, the chief designer of the Soviet space programme was not ready to give up completely on a manned flight to the moon, for Valentin Glushko persisted with dreams for lunar exploration, presenting his last set of ideas in 1986, just three years before his death.

AFTER N-1: A NEW SOVIET MOON PROGRAMME?

Strangely enough, the suspension of the N-1 programme in 1974 did not mean the final end of the Soviet manned moon programme. The new chief designer, Valentin Glushko, announced that the whole space programme would be reappraised and a fresh start made in reconsidering strategic objectives. The only definite decision was that the N-1 would not fly for the time being, if at all. Glushko set up five task forces, one of which was headed by Ivan Prudnikov to develop the idea of a lunar base and another the idea of a new heavy-lift launcher. Glushko personally began to sketch a new series of heavy launchers called the RLAs, or Rocket Launch Apparatus, capable of putting 30, 100 and 200 tonnes into orbit respectively.

When the Politburo met in August 1974, it actually reaffirmed the general objective of Soviet manned missions to the moon. Ivan Prudnikov duly completed, by the end of that year, the plans for a lunar base. The base was called *Zvezda*, or 'star' and featured teams of cosmonauts working on the moon for a year at a time, supplied by the new, proposed heavy-lift rocket. Their proposals were formally tabled, along with the outcome of the four other task forces, in 1975. Design of a heavy-lift launcher appropriately called Vulkan, able to deliver 60 tonnes to lunar orbit, was sketched out. In an abrupt turnaround, Vulkan would be powered with hydrogen fuel, the one system Glushko refused to develop for Korolev. Glushko even designed new hydrogen-fuelled engines, the RD-130 and RD-135, the latter with a specific impulse of no fewer than 450 sec. A lunar expeditionary craft or LEK was designed, not that different from the long-stay lander of Mishin's N1-L3M plan.

Although Glushko put his full force behind *Zvezda*, it attracted little support overall and none from the military at all. Crucially, the president of the Academy of Sciences, Mstislav Keldysh, would not back it. He was never a close friend of Glushko and was wary of the extravagance of the project. The cost, estimated at 100bn roubles, was too much even for a Soviet government not normally shy of extravagant projects. Keldysh let the process of consideration of the project exhaust itself so that it would run out of steam [1]. Glushko tried to save some face with a scaled-down project, but this won little support either. The basic problem was that Glushko had replaced a real rocket (the N-1) and a real programme (N1-L3M), both with diminishing political support, with a theoretical rocket (Vulkan) and a programme (Zvezda) that had none. The Soviet leadership began to regard the Soviet manned moon programme as having been a failure, a waste, a *folie de grandeur* that the country could not afford. Leonid Brezhnev had a mild stroke in 1975 and decisions were taken ever more by a shifting group of ministers and generals. This was not a leadership that would take a big decision and see it through.

In the event, the most significant project to emerge from the strategic reconsideration of 1974–6 was the Energiya–Buran heavy launcher and shuttle system, which was driven by military imperatives to match the American space shuttle. No one can point to a particular day or decision on which the Soviet manned moon programme died, but it withered in mid-1975 and was effectively gone by March the following year, 1976. Despite this, Valentin Glushko even once briefly returned to the moon base idea in the 1980s, outlining how a small base might be built using the Energiya rocket, but he won no support in a country entering ever more difficult economic conditions. Despite their declining political fortunes, the moon base projects reached a certain level of detail and are outlined here.

MOONBASE *GALAKTIKA*, 1969

Moon bases had been part of Soviet thinking for some time. For Glushko, a moon base had a number of attractions. With Apollo over and the shuttle in development, there was no prospect now of the Americans establishing a moon base. By contrast, the world might be impressed by a permanent Soviet settlement on the moon. What would it have looked like?

A considerable amount of homework had already been done on moon bases. Design for a Soviet lunar base dated to the *Galaktika* project, approved by the government in November 1967. This mandated the study of the issues associated with lunar and planetary settlements [2]. The work was done not by one of the normal space design bodies but instead by the bureau associated with the construction of the cosmodromes, Vladimir Barmin's KBOM. Work began in March 1968. Within the broader *Galaktika* programme, whose broad remit was the solar system as a whole, KBOM designed a full lunar base called *Kolumb*, or Columbus, constructed a full-scale habitation model and built a number of scale models, making its report as

Principles of the construction of long-term functioning lunar settlements in late 1969. KBOM designed a moon base for between four and twelve cosmonauts, working on the lunar surface for up to a year at a time. Up to nine modules might be delivered, telescoping out in length after their arrival. The study calculated that establishment of a moon base required the delivery, to the lunar surface, of about 52 tonnes of modules and equipment. Its key elements were:

- Pressurized habitation modules, buried under the regolith for protection from radiation, including a control centre.
- Construction equipment.
- Power supply centre, which could be solar, chemical or nuclear.
- Greenhouse to enrich oxygen, provide food and offer recreation.
- Logistics facilities for oxygen, water, waste disposal.
- Astronomy laboratory.
- Lunar rover, able to carry three cosmonauts across the lunar surface for up to three days to a distance of 250 km.
- Equipment for lunar exploration, such as drills and laboratory devices to examine rocks.

The western edge of the Ocean of Storms, already selected as the prime Soviet manned landing site, was nominated as the best possible location. Barmin was thanked for his work, for which he was paid 50m roubles, but cautioned that it was unlikely to be accomplished until the next century. The existence of this project was not eventually revealed until November 1987, when details were given on the Serbo-Croat and standard Chinese service of Radio Moscow's overseas service.

MOONBASE *ZVEZDA*, 1974

The second moonbase proposal was the *Zvezda* one developed for Valentin Glushko by Ivan Prudnikov in 1974. The crew for the moon base would be brought there by a 31 tonne lunar expeditionary craft, or LEK in Russian. This would use direct ascent, not lunar orbit rendezvous. Once their lunar visit was complete, the three cosmonauts would blast home in their 9.2 tonne upper stage. The reentry vehicle was small, weighing 3.2 tonnes. The initial crew of the base would be three, but this would be doubled as more equipment was ferried up from Earth by Vulkan rockets. The total weight to be transported to the moon would, in the end, be around 130 tonnes, involving up to six Vulkans.

The moon base itself would have three elements: a habitation module, laboratory module and lunar rover. First, there would be a lunar habitation module, or LZhM in Russian. This was a non-returnable 21.5 tonne living and scientific area, 9 m tall, 8 m wide and with a volume of 160 m^3. It would deploy solar panels able to generate 8 kW of electricity. Next was a laboratory production module, the LZM in Russian. Weighing 15.5 tonnes, this would stand 4.5 m tall and have a volume of 100 m^3 for oxygen generation, biotechnology and physics experiments, operated by a single cosmonaut at a time. As was the case with the *Galaktika* proposal, a lunar rover was an essential element. The Lunokhod would measure 4.5 m wide, 8 m long and 3.5 m high, weigh 8.2 tonnes and could transport two cosmonauts up to 200 km distant at a speed of up to 5 km/hr. The rover would be able to drive on expeditions for up to twelve days at a time (a full lunar day), carrying drilling and other scientific equipment.

Valentin Glushko did not give up easily and attempted to resurrect it as *Zvezda II* in the 1980s. It was a scaled-down proposal, using two Energiya rockets rather than the much larger Vulkan [3]. Designed along lines similar to the N1-L3M plan of his deposed rival Vasili Mishin, two Energiyas would place a 74-tonne complex with five cosmonauts on board into lunar orbit. Three would descend to the surface for a twelve-day surface stay. Preliminary designs of the *Zvezda II* mothership and lander were done, both being significantly larger than the LOK and the LK. However, even Glushko must have realized that there was no prospect, at this time, that they would receive serious consideration.

Russia's moon plans

1964–71	N1-L3	Korolev and Mishin
1972–4	N1-L3M	Mishin
1974–6	Vulkan	Glushko

Russia's moon base plans

1967–70	*Galaktika*	Barmin
1974–6	*Zvezda*	Glushko and Prudnikov
1986	*Zvezda II*	Glushko

THE SOVIET/RUSSIAN LUNAR PROGRAMME AFTER 1976

Some time before the cancellation of Luna 25, references to future Soviet lunar exploration had already dried up in the Soviet press. In July 1978 it was briefly reported that a lunar geochemical explorer was under consideration and due to fly by 1983, but nothing more was heard of this project. At around that time, NASA was trying to persuade Congress to fund a lunar geochemical polar orbit – with equal lack of result.

The moon was now relatively well known and Keldysh made the argument to the political leadership that the USSR should no longer try to directly compete with the United States. Both he and the director of the Institute of Space Research (IKI in Russian), Roald Sagdeev, argued that the USSR should concentrate on what it was good at, had proven expertise and did not compete directly with the Americans. This pointed the Soviet Union in only one direction: toward Venus. Here, the Soviet Union had parachuted probes through Venus's atmosphere in 1967 and 1969 (Venera 4, 5–6), soft-landed simple probes on its surface in 1970 and 1972 (Venera 7, 8) and put down sophisticated landers in double missions in 1975, 1978 and 1985 (Venera 9–10, 11–12,

13–14, Vega 1–2). Venera 13 and 14 drilled Venusian soil and analyzed it in an onboard laboratory. Balloons were dropped into the Venusian atmosphere (part of the Vega project). Orbiters first circled the planet in 1975 (Venera 9, 10) and then in 1983 radar-mapped its surface (Venera 15–16). By the end of the Vega programme in 1986, Venus's surface, atmosphere and circumplanetary space had been well characterized.

Mars took second place in the Soviet programme for interplanetary exploration. The Russian Mars 3 probe became the first spacecraft to soft-land on the Red Planet and sent a picture from its surface in December 1971. The Soviet Union obtained a full profile of the atmosphere right down to the surface during the descent of Mars 6 into the Mare Erythraeum in March 1974. After a gap of many years, the USSR went on to organize an imaginative mission to Mars's little moon, Phobos, in 1988–9 (the first probe failed, the second achieved limited success). The Americans began a wave of missions to Mars in the 1990s, each one revealing more and more of what an interesting planet it was.

In the light of the genuine progress made in the successful exploration of Venus and the sustained interest in Mars, it is little wonder that the further scientific exploration of the moon became a low priority. Eventually, though, coinciding with a reforming political leadership in the Soviet Union, some plans were advanced. In 1985, the idea of a lunar polar orbiter was resurrected. In 1987, the Institute for Space Research (IKI) in Moscow gave this mission a target gate of 1993, with a lunar farside sample recovery in 1996 and an unmanned laboratory on the moon, with rovers, in 2000. In its last plan for space development published in 1989 (*The USSR in outer space – the year 2005*), the Soviet Union proposed a lunar polar geophysical orbiter, but few details were given and only a sketchy illustration was published, suggesting it would use the Phobos spacecraft design. At one stage, the project acquired the name *Luna 92*, indicating a 1992 launch date, but it never got beyond the preliminary design stage and the money originally set aside for it was used for the Mars 96 planetary mission instead.

THE REVIVAL? LUNA GLOB

These and other plans were overtaken by political events and the financial crisis that engulfed the Soviet Union and then Russia in the early 1990s. In the post-Soviet space programme, the moon was rarely mentioned. The first instance was in summer 1997, when IKI proposed plans to send a small spacecraft into lunar orbit, using a Molniya rocket from Plesetsk Cosmodrome in northern Russia in 2000 (a Proton would be prohibitively expensive). Over time, this mission acquired the title Luna Glob, or 'lunar globe'. The orbiter would deploy three 250 kg penetrators, modelled on those developed for the Mars 8 mission the previous year. They would dive into the lunar surface at some speed, burrowing seismic and heat flow instruments under the lunar surface, leaving transmitters just above the surface. With small nuclear isotopes, they would transmit for a year, operating as a three-point network to collect information on moonquakes and heat flow. A number of variations on these themes appeared, but none progressed beyond the aspirational stage at this time. The reality was that Russia lacked the financial resources to mount any scientific missions during the 15 years that followed the fall of the Soviet Union and concentrated all its efforts on keeping its manned, military and applications space programmes going. For the time being, the only hopes for more moon probes rested with private industry and Russia providing a booster rocket for a freelance Western venture. One such mission, Trailblazer of the TransOrbital Corporation, was postulated in the early 2000s, using an old Cold War rocket called the Dnepr.

July 2005 saw the Russian Federation announce that there would be a new federal space plan to run through 2015. One of the highlights of the plan was a return to Mars with a new mission to its moon Phobos in 2009, the mission being called Phobos Grunt. Analyzing the plan proved an impossibility, since the government issued only press releases and interviews about it, but never the original text. Almost a year later, it was made known that the plan included, in a revival of the Luna Glob mission, a return to the moon in 2012 [4]. Details of the mission were given by officials of the Russian Space Agency, the Vernadsky Institute and the Institute of Earth Physics. All appeared anxious that Russia, for all its past expertise in the exploration of the moon, should get back in the business of lunar exploration. They were also motivated not just by American plans to return to the moon by 2020 announced by President Bush, but by the prospect of moon probes being sent there by China, India and Japan much sooner.

The new Luna Glob envisaged the launch by a Molniya rocket of a mother ship into lunar orbit. Before arriving at the moon, the mother ship would release a fleet of ten high-speed penetrators to impact into the Sea of Fertility in a circular pattern, each only 2,500 m from the next one, forming a 10-point seismic station. The mother ship would continue into lunar orbit. First, it would deploy two penetrator landers at the Apollo 11 and 12 landing sites, to rebuild the seismic network they began there in 1969. Then it would send a soft-lander down to the south polar region, called the polar station, carrying a seismometer and two spectrometers to detect water ice. The mother ship would act as a relay for the 13 data stations on the lunar surface.

Although the name Dennis Tito will never be as famous or recognizable as many of the great astronaut or cosmonaut heroes, what he did may prove ultimately to be of great importance. When his Soyuz rocket fired him up to the International Space Station in 2001, he became the world's first paying space tourist. In one of the great post-Cold War ironies, commercial space tourism was developed by Russia, albeit by an American company, Space Adventures in cooperation with the builder of Soyuz, the Energiya Corporation. After launching a number of space tourists to the station, Space Adventures decided to offer an even more staggering – and pricey – idea: lunar tourism. Space Adventures' proposal: to offer a six-day loop around the moon in a reconstructed Zond cabin for €80m, with a first flight set for 2009. When the original plans for space tourism were put forward, they were considered a publicity-seeking stunt, but with a queue of millionaires ready to spend the money and go through the year-long training, Space Adventures had established a viable business. Maybe, 40 years later than scheduled, Zond will make a round-the-moon manned flight after all [5].

LOOKING BACK AT THE OLD MOON . . .

How do Soviet space leaders regard their exploration of the moon now? In the early 1990s, the leaders of the space programme at the time emerged from the shadows to tell their story of the moon programme. Inevitably, granted the secrecy of the period, their first concern was to tell *what* happened. Many were directly involved as partisan protagonists, so their comments must be treated cautiously.

One of the first to tell the story was Chief Designer Vasili Mishin, first in magazine articles and then in interviews (his diaries were bought by the Ross Perot Foundation and have yet to be published). He was followed by numerous journalists, writing in dailies such as *Izvestia* and magazines like *Znaniya*. Some of the most detailed information was provided by the head of the cosmonaut squad, General Nikolai Kamanin, who kept a diary throughout the period and which has now been translated [6]. Latest and possibly last of the old guard to speak out was Boris Chertok, who in his eighties compiled a multi-volume memoir, published as *Raketi i Lyudi* (Rockets and people). Regrettably, little has been put in print by the cosmonauts involved in the lunar programme and some of them have already died. Oleg Makarov, for example, although he would have been on the first around-the-moon and first landing mission, was prepared to talk about the planned Zond mission, but would say very little about his own prominent role [7].

All expressed varying levels of regret, even grief, that the Soviet Union failed to win the moon race. Having achieved all the early breakthroughs in space exploration, they took the view that the Soviet Union should, with proper organization, have been able to reach the moon first. 'How could we, after such a bright start, have slipped into second place?' asked military journalist and cosmonaut candidate Col. Mikhail Rebrov [8]. Kamanin told his diary just how difficult it was for him to come to terms with what he regarded, unambiguously and without any mitigating factors, as a crushing defeat: 'We drain the bitter cup of failure to the dregs,' he acidly told his diary. He can't have been the only one.

Most agreed with the reasons advanced by Mishin to explain why the Soviet Union lost: resources much inferior to the United States, the rivalry of the design bureaux, the continual revision and remaking of decisions, the false economy of avoiding comprehensive ground-testing, the death of Korolev at a crucial stage [9]. Most were sympathetic to Mishin, regarding his dismissal and the suspension of the N-1 programme as bad and even unjust decisions. In histories in which blame is liberally apportioned and widely scattered, several focused on Glushko for not being big enough to cooperate with the N-1 project from the start, arguing that he played an inconsistent, spoiling and even vindictive role in the programme [10]. Most felt that the N-1 would probably flown and been a successful rocket, eventually assembling large orbital stations. Chertok believed the Soviet Union did have the capacity to build a proper lunar base in the late 1970s and that such a venture made political, engineering and scientific sense, although it would have been costly [11]. Mishin [12]: *We were able and should have implemented such an expedition after the USA.* 'Only a sense of political embarrassment, out of coming second, after the great rival, pre- vented this from happening,' he said. Most of all he regrets the cancellation of the N-1, the wasted effort, the bitter resentment this caused in the industry and its replacement by an even more expensive programme which was ultimately cancelled in turn. Mishin's final comment: 'We were just a step away from success with the N-1. We could have built a base on the moon by now without stress or hurry.'

Having said this, these accounts are somewhat one-sided. Vladimir Chelomei did not leave memoirs, nor did Valentin Glushko. Although Glushko published technical papers, he never left behind a political statement defending his role in the space programme. When he died in 1989, his vast Energiya bureau was re-divided much as it was before he clustered its constituent companies together in 1974. The original OKB-1, now RKK Energiya, published a vast, colourful company history of the bureau and its projects, providing much of the detail on which an important portion of our knowledge of the Soviet moon programme is based. More critical comments and views come from General Kamanin. A diehard Stalinist, his severest criticisms focused on what he regarded as the poor quality of leadership given by the party and government, his own military and the space programme leadership, like Mishin and Keldysh. He was critical of the N-1 from the start, which he always regarded as an unsuitable and bad rocket: Chelomei's UR-700 would have been better. Patriot though he was, he was overwhelmed in unconditional admiration of America's stunning lunar successes. He resented the way in which they were under-reported and downplayed by the Soviet media and that he could not speak publicly and approvingly of them. He felt just how tough it must be on disappointed Soviet cosmonauts not to fly to the moon. Kamanin was especially critical on how good decision-making was undermined by the corrosive secrecy with which the Soviet lunar programme was run.

Retelling the Soviet side of the moon race, with its setbacks, 'grandiose failures' (Kamanin's words), waste and poor decisions, seems to have given these writers little satisfaction, apart from the unmeasurably important one of making the facts of this hidden history known. They seemed to derive little comfort from the fact that from the chaotic final stages of the moon programme, a plan emerged for the building of space stations. This was a field in which their country became the undisputed world leader and remains so to this day. In his own way, Glushko was vindicated, for in 1987 his replacement for the N-1 did fly, Energiya giving the Soviet Union the most powerful rocket system in the world. Its subsequent cancellation, for economic reasons, can hardly be laid at his door. Unlike the N-1 and more like the Saturn V, the Energiya flew perfectly on its first two testflights and it was not for technical reasons that it never flew again.

Some of the writers refer to the general loss of interest in going to the moon among the Soviet political leadership, now that the Americans had achieved the feat and demonstrated it several times. Afanasayev [13] said that this was the considered view of the Soviet political leadership by 1972 and suggests that the decision to wind down the unmanned lunar programme was taken at around the same time as the decision regarding the manned programme. It is interesting that the political leadership of both the United States and the Soviet Union lost interest in flying to the moon in parallel at around the same time, even though one country had been there and the other had not. The next *grand projet* of Brezhnev's Soviet Union was a more practical, earthly one, the Baikal Amur Railway.

The hardware and rockets from the Soviet lunar programme mostly found their way to museums, like the LK lander. The main collection of unmanned Soviet lunar spacecraft may still be found in the Lavochkin Museum, and that is where Lunokhod 3 may be found. When the financial situation of the Russian space programme reached rock bottom, many of its most famous artefacts were sold, from spacesuits to space cabins, Vasili Mishin's diary, even Sergei Korolev's slide rule. Even real spacecraft were sold. On 11th December 1993, Sotheby's sold Lunokhod for $68,500, but it was explained to the buyer that he would have to collect it from the Sea of Rains!

It is with the moon drivers that we leave the story. Lunokhod was one of the great achievements of the lunar exploration programme, though, as we saw, far from the only one. The moon drivers did everything expected of them and more. Not only had they worked away from home for the year-long journey of Lunokhod and the half-year journey of Lunokhod 2, but they were not even allowed to tell their families or friends where they were or what they were doing. Not until *perestroika* were they allowed to come out of the shadows and tell their remarkable story. Since then, the Lunokhod drivers would gather once a year, on the great 17th November, to recall their experiences in driving on another world. They are older and greyer now and most have now retired. Vyacheslav Dovgan is now a general. Happily, they were at last formally conferred with the medals that they had deserved a quarter century earlier and now wear them with pride [14].

REFERENCES

[1] Sagdeev, Roald Z.: *The making of a Soviet scientist*. John Wiley & Sons, New York, 1994.

[2] Zak, Anatoli: *Manned lunar programme* on his website *www.russianspaceweb.com/lunar*, posted 2002.

[3] Abeelen, Luc van den: The persistent dream – Soviet plans for manned lunar missions. *Journal of the British Interplanetary Society*, vol. 52, April 1999.

[4] Covault, Craig: Russia's lunar return. *Aviation Week and Space Technology*, 5th June 2006.

[5] Jha, Alok: Fly me to the moon – and let me pay among the stars. *The Guardian*, 12th August 2005.

[6] Hendrickx, Bart: The Kamanin diaries, 1960–63. *Journal of the British Interplanetary Society*, vol. 50, #1, January 1997;
  – The Kamanin diaries, 1964–6. *Journal of the British Interplanetary Society*, vol. 51, #11, November 1998;
  – The Kamanin diaries, 1967–8. *Journal of the British Interplanetary Society*, vol. 53, #11/12, November/December 2000;
  – The Kamanin diaries, 1969–71. *Journal of the British Interplanetary Society*, vol. 55, 2002 (referred to collectively as Hendrickx, 1997–2002).

[7] Young, Steven: Soviet Union was far behind in 1960s moon race. *Spaceflight*, vol. 32, #1, January 1990.

[8] Rebrov, Colonel M.: But this is how it was. *Krasnaya Zvezda*, 13th January 1990, translated by Charles E. Noad.

[9] Leskov, Sergei: How we didn't get to the moon. *Izvestia*, 18th August 1989, translated by Charles E. Noad.

[10] Pikul, V.: The history of technology – how we conceded the moon – a look by one of the participants of the N-1 drama and the reasons behind it. *Izobretatel i Ratsionalizator*, #8, August 1990 (in translation).

[11] Smolders, Peter: I met the man who brought the V-2 to Russia. *Spaceflight*, vol. 37, #7, July 1995.

[12] Mishin, Vasili: Why we didn't land on the moon. *Znaniya*, #12, December 1990 (as translated).

[13] Afanasayev, I.B.: Unknown spacecraft. *Znaniya*, #12, December 1991, translated by Ralph Gibbons.

[14] Cirou, Alan: L'histoire secrète des Lunokhod. *Ciel et Espace*, septembre 2004 (avec Jean-René Germain).

9

List of all Soviet moon probes (and related missions)

Date	Name	Type	Outcome
23 Sep 1958		Ye-1	Failed after 90 sec
12 Oct 1958		Ye-1	Failure after 104 sec
4 Dec 1958		Ye-1	Failure
2 Jan 1959	First Cosmic Ship	Ye-1	Passed the moon
18 Jun 1959		Ye-1a	Failure
9 Sep 1959		Ye-1a	Pad abort
12 Sep 1959	Second Cosmic Ship	Ye-1a	Hit the moon
4 Oct 1959	Automatic Interplanetary Station	Ye-2	Circled farside
15 Apr 1960		Ye-2F	Failure
16 Apr 1960		Ye-2F	Failure
4 Jan 1963		Ye-6	Failure
2 Feb 1963		Ye-6	Failure
2 Apr 1963	Luna 4	Ye-6	Missed moon
21 Mar 1964		Ye-6	Failure
12 Mar 1965	Cosmos 60	Ye-6	Failure
10 Apr 1965		Ye-6	Failure
9 May 1965	Luna 5	Ye-6	Crashed on moon
8 Jun 1965	Luna 6	Ye-6	Missed moon
18 Jul 1965	Zond 3	3MV	Passed, imaged moon
4 Oct 1965	Luna 7	Ye-6	Crashed on moon
3 Dec 1965	Luna 8	Ye-6	Crashed on moon
31 Jan 1966	Luna 9	Ye-6M	Soft-landed
1 Mar 1966	Cosmos 111	Ye-6S	Failure
31 Mar 1966	Luna 10	Ye-6S	Orbited moon
24 Aug 1966	Luna 11	Ye-6LF	Orbited moon

Date	Name	Type	Outcome
22 Oct 1966	Luna 12	Ye-6LF	Orbited moon
21 Dec 1966	Luna 13	Ye-6M	Soft-landed
10 Mar 1967	Cosmos 146	L-1	High-altitude test
8 Apr 1967	Cosmos 154	L-1	
17 May 1967	Cosmos 159	Ye-6LS	Failure
28 Sep 1967		L-1	Failure
23 Nov 1967		L-1	Failure
4 Mar 1968	Zond 4	L-1	
7 Feb 1968		Ye-6LS	Failure
7 Apr 1968	Luna 14	Ye-6LS	Orbited moon
23 Apr 1968		L-1	Failure
15 Sep 1968	Zond 5	L-1	Returned to the Earth
14 Nov 1968	Zond 6	L-1	Returned to the Earth
20 Jan 1969		L-1	Failure
19 Feb 1969		Ye-8	Failure
21 Feb 1969	N-1	L-1S	All-up test, failure
14 Jun 1969		Ye-8-5	Failure
3 Jul 1969		L-1S	All-up test, failure
13 Jul 1969	Luna 15	Ye-8-5	Crashed on landing
8 Aug 1969	Zond 7	L-1	Returned to the Earth
23 Sep 1969	Cosmos 300	Ye-8-5	Failure
22 Oct 1969	Cosmos 305	Ye-8-5	Failure
18 Nov 1969		KL-1E	Failure
19 Feb 1970		Ye-8-5	Failure
12 Sep 1970	Luna 16	Ye-8-5	Returned samples
20 Oct 1970	Zond 8	L-1	Returned to the Earth
10 Nov 1970	Luna 17/Lunokhod	Ye-8	Landed rover
24 Nov 1970	Cosmos 379	LK	LK test
2 Dec 1970	Cosmos 382	KL-1E	Block D test
26 Feb 1971	Cosmos 398	LK	LK test
27 Jun 1971	N-1	[LK, LOK]	Failure
12 Aug 1971	Cosmos 434	LK	LK test
2 Sep 1971	Luna 18	Ye-8-5	Crashed on landing
28 Sep 1971	Luna 19	Ye-8LS	Orbited the moon
14 Feb 1972	Luna 20	Ye-8-5	Returned samples
23 Nov 1972	N-1	LOK	All-up, with LOK (failure)
8 Jan 1973	Luna 21/Lunokhod 2	Ye-8	Landed rover
2 Jun 1974	Luna 22	Ye-8LS	Orbited the moon
28 Oct 1974	Luna 23 (failure)	Ye-8-5M	Crashed on landing
16 Oct 1975		Ye-8-5M	Failure
9 Aug 1976	Luna 24	Ye-8-5M	Returned samples

WHERE ARE THEY NOW? LOCATION OF SOVIET MOON PROBES

These are the current locations of Soviet moon probes:

In solar orbit

Name	Date of lunar flyby	Distance from moon (km)
First Cosmic Ship	4 Jan 1959	5,965
Luna 6	11 June 1965	160,935
Zond 3	20 July 1965	9,219

In eccentric Earth orbit

Name	Date of lunar flyby	Distance from moon (km)	Final orbit (km)
Luna 4	5 Apr 1963	8,451	89,250–694,000*

* There are reports that Luna 4 was eventually perturbed into solar orbit.

Impacted on the moon's surface

Name	Date of impact	Coordinates	Location
Second Cosmic Ship	14 Sep 1959	39°N, 1°W	Marsh of Decay*
Luna 5	12 May 1965	31°S, 8°W	Sea of Clouds
Luna 7	7 Oct 1965	9.8°N, 47.8°W	Kepler, Ocean of Storms
Luna 8	6 Dec 1965	9.8°N, 63.3°W	Ocean of Storms
Luna 15	21 Jul 1969	17°N, 60°E	Sea of Crises
Luna 18	10 Sep 1971	56.5°E, 3.57°N	Apollonius
Luna 23	6 Nov 1974	13°N, 62°E	Sea of Crises

*Also its upper stage, place of impact not known.

On the moon's surface, intact

Name	Date of arrival	Coordinates	Location
Luna 9	2 Feb 1966	64.37°W, 7.08°N	Ocean of Storms
Luna 13	24 Dec 1966	18.87°N, 62.05°W	Ocean of Storms
Luna 16 landing stage	20 Sep 1970	0.68°S, 56.3°E	Sea of Fertility
Luna 17 landing stage	17 Nov 1970	38.28°N, 35°W	Bay of Rains (1)
Luna 20 landing stage	21 Feb 1972	3.53°N, 56.55°E	Apollonius
Luna 21 landing stage	15 Jan 1973	25.9°N, 30.5°E	Le Monnier Crater (2)
Luna 24 landing stage	18 Aug 1976	12.8°N, 62.2°E	Sea of Crises

(1) Also Lunokhod; (2) also Lunokhod 2.

In lunar orbit

Name	Date of arrival	Equatorial plane
Luna 10	3 Apr 1966	71.9°
Luna 11	27 Aug 1966	27°
Luna 12	25 Oct 1966	15°
Luna 14	10 Apr 1968	42°
Luna 19	5 Oct 1971	40°
Luna 22	6 Jun 1974	19°, then 21°

Returned to Earth from the moon's surface

Name	Recovery	Landing location on Earth
Luna 16	24 Sep 1970	80 km SE of Dzhezhkazgan, 47.4°N, 68.6°E
Luna 20	26 Feb 1972	Kazakhstan, 48°N, 67.56°E
Luna 24	21 Aug 1976	Surgut, Siberia, 61.06°N, 75.9°E

Returned to Earth after circling the moon

Name	Recovery	Distance over moon (km)	Landing location
Zond 5	14 Sep 1968	1,950	Indian Ocean, 32°38′S, 65°33′E
Zond 6	17 Nov 1968	2,420	Kazakhstan
Zond 7	8 Aug 1969	2,000	Kazakhstan
Zond 8	20 Oct 1970	1,100	Indian Ocean, 730 km SE Chagos

The Automatic Interplanetary Station passed the moon at a distance of 6,200 km on 6th October 1959 and returned to the vicinity of the Earth, but no attempt was made at recovery.

Bibliographical note and bibliography

BIBLIOGRAPHICAL NOTE

Any book on Soviet and Russian lunar exploration has to face problems of information sources and their reliability. Even such apparently mundane and non-controversial matters as the paths taken by Soviet spacecraft as they circled the moon and the precise coordinates as to where they landed can be problematic, with official sources quoting different and even contradictory details – and then revising them! During the peak of the moon race, the official organs of Soviet government issued an economy of information on certain spacecraft and even disinformation on others. There is no official, comprehensive authorized history of the Soviet moon programme, which makes the assembly of the story all the more challenging, interesting and necessary. I have tried to put together the most accurate sources that best fit the known facts, 'the best version of the truth available' – but over time these will be superseded as new information sources become available. The story of the Soviet/Russian moon programme is still, as the saying goes now, a 'site under construction'.

A book such as this invariably relies on a variety of diverse sources. Some of the main elements are outlined here. First, there has long been a Western tradition of analyzing the Soviet lunar programme, putting together the best possible version of the truth available from official statements, Western intelligence analysis and an examination of trajectories and orbits. Here, Stoiko (1970) and Gatland (1972) were the pioneers, both giving due prominence to the lunar programme. They were followed by Clark (1988–2005) who has made multiple, penetrating, in-depth analyses of the performance of Soviet spacecraft and has invariably been vindicated by the official story emerging years later. Their work has been supplemented by specialized studies such as those of: Vick in analysing Russian rockets and launch facilities (1994–6); Rex Hall, who identified the members of the cosmonaut squad and their roles

(1988–2003); Gordon Hooper (1990), who assembled their biographies; and Jim Harford (1997), who penned the authoritative biography of Sergei Korolev. Others have brought different knowledge to bear – for example, in the area of tracking (Sven Grahn); the analysis of hardware (David Portree, 1995; Nicholas Johnson, 1994); the performance of rockets (Berry Sanders, 1996–7); and the development of space equipment (Don P. Mitchell). Mark Wade and Anatoli Zak have done much to assemble what is now known of the Soviet moon programme and make it globally and readily available on the Internet to amateurs, professionals and historians alike. Recently, Pesavento and Vick (2004) wrote a lengthy heretical series in the historical magazine *Quest*, re-opening the debate about Soviet lunar capabilities and intentions during the pivotal years 1968–9.

Following Soviet accounts of their early lunar programme required a challenging effort to separate the respective strands of reporting, science, human interest, engineering and achievement, news management and even disinformation. Soviet lunar missions were publicized in standard English language outlets, such as Radio Moscow's World Service, magazines, periodical and miscellaneous grey literature (e.g., *Science and Life*, *Soviet Weekly*, *Sputnik*, *Soviet booklet* series). These were all used where they were available. Scientific outcomes were published in a number of specialized international journals.

The precise nature of the Soviet lunar effort did not become clear until a number of designers, scientists and journalists were given or took the opportunity to speak more openly about the Soviet side of the moon race. Most prominent of these was Chief Designer Vasili Mishin (1990), but he was accompanied by a number of scientists, journalists and colleagues, such as Leskov (1989), Chernyshov (1990), Rebrov (1990), Filin (1991), Afanasayev (1991) and Lebedev (1992). On its 50th anniversary, in 1996, the Energiya design bureau published its official history, full of hitherto unknown details of its moon programme. Russian journalists and space enthusiasts have now been able to tell the story of their country's space programme. In a detailed 13-part series, Varfolomeyev (1995–2002) has reconstructed, for *Spaceflight*, the technical history of many of the key rocket programmes of the period. Perhaps the most remarkable contemporaneous document from the period was the diary of the head of the Soviet cosmonaut squad, General Nikolai Kamanin, whose record has been painstakingly and faithfully reconstructed by Hendrickx (1997–2002), whose endeavours in translating and interpretation are an enduring contribution to history. Cosmonauts (e.g., Alexei Leonov) and designers (e.g., Chertok) have now written memoirs. Soviet historical documentation from the period has now become more widely available and here Siddiqi (2000) has made the most impressively scholarly interpretation, one likely to be the principal point of reference for many years.

As the generation that managed the Soviet lunar programme begins to pass on, the preservation of that record becomes more important. In recent times, writers such as Yuri Surkov (1997) have now come to publish the scientific results of Russian lunar and planetary exploration. The first attempt to assemble a web-based inventory of Soviet lunar science was undertaken by the American space agency, NASA, where the Goddard Spaceflight Centre began to put together an archive of Soviet lunar and planetary science which was made available on the Internet in the NSSDC *Master catalogue*.

BIBLIOGRAPHY

Abeelen, Luc van den: Soviet lunar landing programme. *Spaceflight*, vol. 36, #3, March 1994.

Abeelen, Luc van den: The persistent dream – Soviet plans for manned lunar missions. *Journal of the British Interplanetary Society*, vol. 52, April 1999.

Abramov, Isaac P. and Skoog, A. Ingemaar: *Russian spacesuits*. Springer/Praxis, Chichester, UK, 2003.

Afanasayev, I.B.: Unknown spacecraft. *Znaniya*, #12, December 1991, translated by Ralph Gibbons.

Ball, Andrew: *Automatic Interplanetary Stations*. Paper given to British Interplanetary Society, 7th June 2003.

British Broadcasting Corporation: *What if?* Broadcast, Radio 4, 3rd April 2003.

Burchett, Wilfred and Purdy, Anthony: *Cosmonaut Yuri Gagarin – first man in space*. Panther, London, 1961.

Caidin, Martin: *Race for the moon*. Kimber, London, 1959.

Carrier, W. David III: *Soviet rover systems*. Paper presented at Space programmes and technology conference, American Institute of Aeronautics and Astronautics, Huntsville, AL, 24th–26th March 1992. Lunar Geotechnical Institute, Lakeland, FL.

Chaikin, Andrew: The other moon landings. *Air and Space*, vol. 18, #6, February/March 2004.

Chernyshov, M: Why were Soviet cosmonauts not on the moon? *Leninskoye Znamya*, 1st August 1990 (as translated).

Cirou, Alan: L'histoire secrète des Lunokhod. *Ciel et Espace*, septembre 2004 (avec Jean-René Germain).

Clark, Phil: *The Soviet manned space programme*. Salamander, London, 1988.

Clark, Phillip S.: The Soviet manned lunar programme and its legacy. *Space policy*, August 1991.

Clark, Phillip S.: Obscure unmanned Soviet satellite missions. *Journal of the British Interplanetary Society*, vol. 46, #10, October 1993.

Clark, Phillip S. and Gibbons, Ralph: The evolution of the Soyuz programme. *Journal of the British Interplanetary Society*, vol. 46, #10, October 1993.

Clark, Phillip S.: The history and projects of the Yuzhnoye design bureau. *Journal of the British Interplanetary Society*, vol. 49, #7, July 1996.

Clark, Phil: Analysis of Soviet lunar missions. *Space Chronicle, Journal of the British Interplanetary Society*, vol. 57, supplement 1, 2004.

Clark, Phillip S.: Masses of Soviet Luna spacecraft. *Space Chronicle, Journal of the British Interplanetary Society*, vol. 58, supplement 2, 2005.

Covault, Craig: Russia's lunar return. *Aviation Week and Space Technology*, 5th June 2006.

Da Costa, Neil: *Visit to Kaliningrad*. Paper presented to British Interplanetary Society, 5th June 1999.

Designer Mishin speaks on early Soviet space programmes and the manned lunar project. *Spaceflight*, vol. 32, #3, March 1990.

Filin, V: Development of lunar spacecraft for manned lunar landing programme. *Aviatsiya i Kosmonautika*, #12, December 1991 (as translated).

Gatland, Kenneth: *Robot explorers*. Blandford, London, 1972.

Gorin, Peter A.: Rising from the cradle – Soviet public perceptions of space flight before Sputnik, in Roger Launius, John Logsdon and Robert Smith (eds): *Reconsidering Sputnik – forty years since the Soviet satellite*. Harwood Academic, Amsterdam, 2000.

Grahn, Sven:
- Mission profiles of early Soviet lunar probes;
- Radio systems used by the Luna 15–24 series of spacecraft;
- Tracking Luna 24 from Florida and Sweden;
- Reception of signals on 183.54 MHz from the Luna 20 return spacecraft in Stockholm;
- Why the west did not believe in Luna 1;
- Zond 7K-L-1 cockpit layout;
- The radio systems of the Luna 4–14 series;
- The radio systems of the early Luna probes;
- Soviet/Russian OKIK ground station sites;
- The Soviet/Russian deep space network;
- Jodrell Bank's role in early space tracking;
- The Kontakt rendezvous and docking system: *www.users.wineasy.se/svengrahn/histind*

Grahn, Sven and Flagg, Richard S.: *Mission profiles of 7K L-1 flights, www.users.wineasy.se/svengrahn/histind*, 2000.

Gracieux, Serge: Le joker Soviétique. *Ciel et Espace*, mai/juin 2005.

Haessler, Dietrich: Soviet rocket motors on view. *Spaceflight*, vol. 35, #2, February 1993.

Hall, Rex: The Soviet cosmonaut team. *Journal of the British Interplanetary Society*, vol. 41, #3, March 1988.

Hall, Rex: Civilians in the cosmonaut team. *Journal of the British Interplanetary Society*, vol. 46, #10. October 1993.

Hall, Rex D. and Shayler, David J.: *Soyuz – a universal spacecraft*. Springer/Praxis, Chichester, UK, 2003

Harford, Jim: Korolev – how one man masterminded the Soviet drive to beat America to the moon. *John Wiley & Sons*, New York, 1997.

Hendrickx, Bart:
- The Kamanin diaries, 1960–63. *Journal of the British Interplanetary Society*, vol 50, #1, January 1997;
- The Kamanin diaries, 1964–6. *Journal of the British Interplanetary Society*, vol. 51, #11, November 1998;
- The Kamanin diaries, 1967–8. *Journal of the British Interplanetary Society*, vol. 53, #11/12, November/December 2000;
- The Kamanin diaries, 1969–71. *Journal of the British Interplanetary Society*, vol. 55, 2002 (referred to collectively as Hendrickx, 1997–2002).

Hendrickx, Bart: The origins and evolution of the Energiya rocket family. *Journal of the British Interplanetary Society*, vol. 55, #7/8, July/August 2002.

Hooper, Gordon: *The Soviet cosmonaut team* (2 vols, 2nd edition). GRH Publications, Lowestoft, UK, 1990.

Huntress, W.T., Moroz, V.I. and Shevalev, I.L.: Lunar and robotic exploration missions in the 20th century. *Space Science Review*, vol. 107, 2003.

Institute of Space Research, USSR Academy of Sciences: *The Soviet programme of space exploration for the period beginning in the year 2000: plans, projects and international cooperation, part 2: the planets and small planets of the solar system*. Moscow, 1987.

Ivanovsky, Oleg: Memoir, in John Rhea (ed.): *Roads to space – an oral history of the Soviet space programme*. McGraw-Hill, London, 1995.

Ivashkin, V.V.: *On the history of space navigation development*. American Astronautical Society, *History* series, vol. 22, 1993.

Jha, Alok: Fly me to the moon – and let me pay among the stars. *The Guardian*, 12th August 2005.

Johnson, Nicholas: *The Soviet reach for the moon – the L-1 and L-3 manned lunar programs and the story of the N-1 moon rocket*. Cosmos Books, Washington DC, 1994.

Kemurdzhian, A.L., Gromov, V.V., Kazhakalo, I.F., Kozlov, G.V., Komissarov, V.I., Korepanov, G.N., Martinov, B.N., Malenkov, V.I., Mityskevich, K.V., Mishkinyuk, V.K. *et al.*: Soviet developments of planetary rovers 1964–1990. CNES & Editions Cepadues: *Missions, technologies and design of planetary mobile vehicles, 1993*, Proceedings of conference, Toulouse, France, September 1992.

Khrushchev, Sergei: The first Earth satellite – a retrospective view from the future, in Roger Launius, John Logsdon and Robert Smith (eds): *Reconsidering Sputnik – forty years since the Soviet satellite*. Harwood Academic, Amsterdam, 2000.

Lebedev, D.A.: The N1-L3 programme. *Spaceflight*, vol. 34, #9, September 1992.

Leonov, Alexei and Scott, David: *Two sides of the moon – our story of the cold war space race*. Simon & Schuster, London, 2004.

Leskov, Sergei: How we didn't get to the moon. *Izvestia*, 18th August 1989, translated by Charles E Noad.

Lipsky, Yuri: Major victory for Soviet science – new data on the invisible side of the moon. *Pravda*, 17th August 1965.

Lovell, Bernard:
   – *The story of Jodrell Bank*. Oxford University Press, London, 1968;
   – *Out of the zenith – Jodrell Bank, 1957–70*. Oxford University Press, London, 1973.

Luna 12 transmits. *Pravda*, 6th November 1966, as translated by NASA.

Marinin, Igor and Lissov, Igor: Russian scientist cosmonauts – raw deal for real science in space. *Spaceflight*, vol. 38, #11, November 1996.

Minikin, S.N. and Ulubekov, A.T.: *Earth–space–moon*. Mashinostroeniye Press, Moscow, 1972.

Matson, Wayne R: *Cosmonautics – a colourful history*. Cosmos Books, Washington DC, 1994.

Mikhailov, A.A.: *On the reverse side of the moon*. Paper presented to the XI International Astronautical Conference, Stockholm, 1960.

Mills, Phil: Aspects of the Soyuz 7K-LOK Luniy Orbital Korabl lunar orbital spaceship. *Space Chronicle. Journal of the British Interplanetary Society*, Vol. 57, supplement 1, 2004.

Mitchell, Don P. (2003–4):
   – Soviet interplanetary propulsion systems;
   – Inventing the interplanetary probe;
   – Soviet space cameras;
   – Soviet telemetry systems;
   – Remote scientific sensors: *www.mentallandscape.com*

Moon programme that faltered – Vasili Mishin outlines Soviet manned lunar project. *Spaceflight*, vol. 33, #1, January 1991.

Mosnews: Soviet scientists planned 'invulnerable' military headquarters on the moon. *Mosnews*, 20th September 2004.

Mishin, Vasili: Why we didn't land on the moon. *Znaniya*, #12, December 1990 (as translated).

Morring, Frank: Moon mapper; 'Touch the water'. *Aviation Week and Space Technology*, 23rd January 2006.

Nesmyanov, A: Soviet moon rockets – a report on the flight and scientific results of the second and third space rockets. *Soviet booklet* series #62, London, 1960.

Pesavento, Peter: Soviet space programme – CIA documents reveal new historical information. *Spaceflight*, vol. 35, #7, July 1993.

Pesavento, Peter: Soviet circumlunar programme hardware revealed. *Spaceflight*, vol. 36, #11, November 1994.

Pesavento, Peter and Vick, Charles P.: The moon race end game – a new assessment of Soviet crewed aspirations. *Quest*, vol. 11, #1, #2, 2004 (in two parts).

Petrovich, G.V.: Some problems of the future exploration of the moon with rockets, appendix to Nesmyanov, A.: Soviet moon rockets – a report on the flight and scientific results of the second and third space rockets. *Soviet booklet* series #62, London, 1960.

Pikul, V.: The history of technology – how we conceded the moon – a look by one of the participants of the N-1 drama and the reasons behind it. *Izobretatel i Ratsionalizator*, #8, August 1990 (in translation).

Pirard, Theo: The cosmonauts missed the moon! *Spaceflight*, vol. 35, #12, December 1993.

Pomeroy, John H. (ed.): *Soviet–American conference on the geochemistry of the moon and planets.* NASA, Washington DC, 1977, in two parts.

Portree, David S.F.: *Mir hardware heritage.* NASA, Houston, TX, 1995.

Raushenbakh, Boris: The Soviet programme of moon surface exploration, 1966–79. American Astronautical Society, *History* series, vol. 23, 1994.

Rebrov, M.: But this is how it was. *Krasnaya Zvezda*, 13th January 1990, translated by Charles E. Noad.

Rhea, John (ed.): *Roads to space – an oral history of the Soviet space programme.* McGraw-Hill, London, 1995.

RKK Energiya: *The legacy of SP Korolev.* Apogee Books, Burlington, Ontario, with RKK Energiya, Moscow, 2001.

Sagdeev, Roald Z.: *The making of a Soviet scientist.* John Wiley & Sons, New York, 1994.

Salakhutdinov, G.: Once more about space – interview with Vasili Mishin. *Ogonek*, #34, 18th–25th August 1990 (in translation).

Sanders, Berry: An analysis of the trajectory and performance of the N-1 lunar launch vehicle. *Journal of the British Interplanetary Society*, vol. 49, #7, July 1996.

Sanders, Berry: An updated analysis of the three stage N-1 lunar launch vehicle. *Journal of the British Interplanetary Society*, vol. 50, #8, August 1997.

Shayler, David J.: *Space suits.* Presentation to the British Interplanetary Society, 3rd June 1989.

Shevchenko, V.V.: Mare Moskvi. *Science and Life*, vol. 3, #88.

Sheldon, Charles: *Soviet space programmes, 1976–1980, unmanned space activities*, part 3. 90th Congress of the United States, US Government Printing Office, Washington DC, 1985.

Siddiqi, Asif: Early satellite studies in the Soviet Union, 1947–57, Part 2. *Spaceflight*, vol. 39, #11, November 1997.

Siddiqi, Asif: The decision to go to the moon. *Spaceflight*:
 – vol. 40, #5, May 1998 (part 1);
 – vol. 40, #6, June 1998 (part 2).

Siddiqi, A.: First to the moon. *Journal of the British Interplanetary Society*, vol. 51, #6, June 1998, Soviet/CIS *Astronautics* series, part 14. See also comments on the paper by Timothy Varfolomeyev, *Journal of the British Interplanetary Society*, vol. 52, #4, April 1999, 157–161.

Siddiqi, Asif: *The challenge to Apollo.* NASA, Washington DC, 2000.

Siddiqi, Asif A.: Rocket engines from the Glushko design bureau. *Journal of the British Interplanetary Society*, vol. 54, #9/10, 2001.

Siddiqi, Asif, Hendrickx, Bart and Varfolomeyev, Timothy: *The tough road travelled: a new look at the second generation Luna probes*. Unpublished paper for Journal of the British Interplanetary Society.

Siddiqi, Asif: *Deep space chronicle*. NASA, Washington DC, 2001.

Smith, Andrew: *Moondust – in search of the men who fell to Earth*. Bloomsbury, London, 2005.

Smolders, Peter: I met the man who brought the V-2 to Russia. *Spaceflight*, vol. 37, #7, July 1995.

Sokolov, Oleg: The race to the moon – a look back from Baikonour. American Astronautical Society, *History* series, vol. 23, 1994.

Sokolov, Oleg: Realized and non-realized projects in the Soviet manned lunar programme. American Astronautical Society, *History* series, vol. 25, 1996.

Stoiko, Michael (1970): *Soviet rocketry – the first decade of achievement*. David & Charles, Newton Abbot, UK.

Surkov, Yuri: *Exploration of terrestrial planets from spacecraft – instrumentation, investigation, interpretation*, 2nd edition. Wiley/Praxis, Chichester, UK, 1997.

Tyulin, Georgi: Memoirs, in John Rhea (ed.): *Roads to space – an oral history of the Soviet space programme*. McGraw-Hill, London, 1995.

Ulivi, Paolo: *Moon exploration – an engineering history*. Springer-Verlag, London, 2003.

United States Congress: *Soviet space programs, 1976–80: unmanned space activities*. Washington DC, 1985, 99th Congress.

Varfolomeyev, Timothy: Soviet rocketry that conquered space. *Spaceflight*, in 13 parts:

    1 Vol. 37, #8, August 1995;
    2 Vol. 38, #2, February 1996;
    3 Vol. 38, #6, June 1996;
    4 Vol. 40, #1, January 1998;
    5 Vol. 40, #3, March 1998;
    6 Vol. 40, #5, May 1998;
    7 Vol. 40, #9, September 1998;
    8 Vol. 40, #12, December 1998;
    9 Vol. 41, #5, May 1999;
    10 Vol. 42, #4, April 2000;
    11 Vol. 42, #10 October 2000;
    12 Vol. 43, #1, January 2001;
    13 Vol. 43 #4 April 2001 (referred to as Varfolomeyev, 1995–2001)

Vasilyev, V.: Drilling in the lunar highlands. *Nedelya*, 21st–27th February 1972.

Vick, Charles P.: The Mishin mission, December 1962–December 1993. *Journal of the British Interplanetary Society*, vol. 47, #9, September 1994.

Vick, Charles P.: Launch site infrastructure – CIA declassifies N-1/L-3 details. *Spaceflight*, vol. 38, #1, January 1996.

Vick, Charles P.: Korolev's lunar mission profile. *Spaceflight*, vol. 38, #8, August 1996.

Wachtel, Claude: Design studies of the Vostok Zh and Soyuz spacecraft. *Journal of the British Interplanetary Society*, vol. 35, 1982.

Wade, Mark: *Encyclopaedia Astronautica*, www.astronautix.com

Wade, Mark: *Energiya – the decision*, www.astronautix.com, 2000

Wade, Mark: *Soyuz 7K-LOK*, www.astronautix.com, 2003.

Wilson, Andrew: *Solar system log*. Jane's.

Wotzlaw, Stefan, Käsmann, Ferdinand and Nagel, Michael: Proton – development of a Russian launch vehicle. *Journal of the British Interplanetary Society*, vol. 51, #1, January 1998.

Wright, Pearce: Vasili Mishin – space boss scapegoated for failure to put a man on the moon. *The Guardian*, 1st November 2001.

Young, Steven: Soviet Union was far behind in 1960s moon race. *Spaceflight*, vol. 32, #1, January 1990.

Zak, Anatoli: Manned Martian expedition, *www.russianspaceweb.com*, 2001.

Zak, Anatoli: Manned lunar programme, *www.russianspaceweb.com/lunar*, posted 2002.

Index